E-WORLD
EMERGING TRENDS IN INFORMATION TECHNOLOGY

E-WORLD
EMERGING TRENDS IN INFORMATION TECHNOLOGY

Rohan Sharma

RANDOM PUBLICATIONS

NEW DELHI - 110 002 (INDIA)

E-World: Emerging Trends in Information Technology

ISBN 978-93-51116-75-2

Published in 2015 in India by

Reprint 2019

RANDOM PUBLICATIONS

4376-A/4B, Gali Murari Lal, Ansari Road

New Delhi-110 002

Phone: +9111-43580356, 23289044

E-mail: randomexports@gmail.com; sales@randompublications.com; info@randompublications.com

Type Setting by: Friends Media, Delhi-110089

Printed at : Mehra Printers, Delhi-110092

Preface

21st century has been defined by application of and advancement in information technology. Information technology has become an integral part of our daily life. According to Information Technology Association of America, information technology is defined as "the study, design, development, application, implementation, support or management of computer-based information systems." In the history of civilization, no work of science has so comprehensively impacted on the course of human development as IT. It has succeeded in breaking old barriers and building new interconnections in the emerging global scenario. IT has become the chief determinant of the progress of nations, communities and individuals. Information technology has served as a big change agent in different aspect of business and society. It has proven game changer in resolving economic and social issues. Advancement and application of information technology are ever changing. One of the most talked about concept in information technology is the cloud computing. Clouding computing is defined as utilization of computing services, i.e. software as well as hardware as a service over a network. Typically, this network is the internet. Cloud computing offers three types of broad services mainly Infrastructure as a Service (IaaS), Platform as a Service (PaaS) and Software as a Service (SaaS). Another emerging trend within information technology is mobile applications.

Mobile application or mobile app has become a success since its introduction. They are designed to run on Smart phone, tablets and other mobile devices. They are available as a download from various mobile operating systems like Apple, Blackberry, Nokia, etc. Some of the mobile app are available free where as some involve download cost. The revenue collected is shared between app distributor and app developer. User interface has undergone a revolution since introduction of touch screen. The touch screen capability has revolutionized way end users interact with application. Touch screen enables the user to directly interact with what is displayed and also removes any intermediate hand-held device like the mouse. The field of analytics has grown many folds in recent years. Analytics is a process which helps in discovering the informational patterns with data. The field of analytics is a combination of statistics, computer programming and operations research. The every changing field of information technology has seen great advancement and changes in the last decade. And from the emerging trend, it can be concluded that its influence on business is ever growing, and it will help companies to serve customers better.

This book provides a broad coverage of many advanced concepts in the field of information technology. This comprehensive book will be an essential resource book for students and IT professionals.

I thank all members of my team who have helped in the preparation of the book. My special thanks go to "Random Publications" who have published the book.

— *Rohan Sharma*

Contents

Chapter 1

Introduction

Information technology (IT), as defined by the Information Technology Association of America (ITAA) is: "the study, design, development, implementation, support or management of computer-based information systems, particularly software applications and computer hardware." In short, IT deals with the use of electronic computers and computer software to convert, store, protect, process, transmit and retrieve information, securely.

In this definition, the term "information" can usually be replaced by "data" without loss of meaning. Recently it has become popular to broaden the term to explicitly include the field of electronic communication so that people tend to use the abbreviation ICT (Information and Communication Technology). Strictly speaking, this name contains some redundancy. The World has entered the new millennium, which is going to be an Information Technology Age. Today Computers have not only assumed strategic importance in the corporate world, they are being effectively used in other fields ranging from space exploration to food processing and banking to communication etc.

In this era of Information Technology, which has revolutionised the whole world, INDIA has stood to the world standards and is being regarded the World over for it's skilled IT Professionals. Even the government has recognised the promising future of this industry and has formed a new IT Ministry which will give a boost to this industry.

Presence of Multinationals like IBM, HP has made possible transfer of hardware technology into the country, Internet has further bridged the time gap ensuring arrival of the latest technology simultaneously in INDIA and the West without delay. The domestic hardware industry has witnessed quantum growth in the turnover and profits, which is largely attributable to our liberalized economy. On the software front various software giants like IBM, Motorola, Oracle, Samsung, HP, Digital, Unisys, AT&T, ICL, Fugitsu, etc. have opened software development centres in the country.

The number of professionals in this industry is believed to have crossed the 2 lakh mark and still there is a huge gap between the demand and supply of professionals which is an encouraging sign. The US alone requires around 3 lakh professionals, leave aside other countries. It is believed around 65% of the world's software is produced in INDIA. INDIA is also exporting software to around 100 countries including many countries which are exclusive buyers from INDIA. Many Indian IT companies like Infosys, Wipro, NIIT, Zenith computers, Satyam Computers, STG, Pentafore Software, Mastek etc. are expanding themselves in a big way making their presence felt globally.

The phenomenal growth which this industry is witnessing has lead to it being recognised as one of the highly paid industry. As this industry is young, the average age of professionals is lower than any other industry. The availability of skilled, qualified professionals is hardly able to meet the requirement, leaving a large gap to be filled by those with ambition, aptitude and willingness to work hard.

Following are the career areas in this industry.

1. *Software:* Software is a set of programmed instructions that enable the computer to perform specified functions. This industry offers vast range of employment opportunities in fields starting from data entry, computer operations, programming, system analysis to system designing, system engineering and operation management. Today, this industry is one of the most sought after industry as far as returns are concerned.
2. *Hardware:* Hardware denotes the physical components of a computer. The key areas which can be taken up as career includes hardware designing, research and development, assembling, manufacturing and maintenance of computer components.
3. *Telecom:* Telecommunications is a growing industry and has created ample of scope for people joining telecommunication engineering. The various fields which can be pursued as career in this industry include voice processing units, telephone technology, wireless technology, cellular technology, touchtone telephones, microwave and satellite communication systems.
4. *Sales & Marketing:* Sales and marketing in IT industry is a specialist task and requires a thorough study of products and

their competition. Marketing covers selling of both hardware and software products. This job is both competetive and challenging but not difficult as the demand for hardware and software is quite high and does not require much effort in fixing a deal.

5. *E-commerce and Web Development:* The coming together of computers and telecommunication technologies has led to the emergence of electronic commerce. Marketing and banking activities have already started taking place worldwide through Internet generating an Internet economy. Most of the future jobs anticipated in IT industry will be Internet related. These will include consumer/buyer information and services, developing and managing Web-portals, creation of Internet market, customer management, consulting etc.
6. *Enterprise Resource Planning:* ERP is planning the resources of entire network of an enterprise. Various fields like purchase, production, marketing, distribution, sales, service, inventory, finance, accounts, human resource are integrated on a single software through ERP, to facilitate better coordination. Professionals with engineering background in mechanical, civil, manufacturing, textile as well as MBAs, chartered accountants, cost accountants can join technical ERP.
7. *Computer Operator:* A computer operator's job is to run both the computer and its peripheral equipment. The job may vary depending upon size of the organisation and its exposure to computers. An operator usually works on instructions issued by programmers or operation managers. Upon acquiring programming training, he can even end up being a full time efficient application programmer.
8. *Data Entry*: Data entry is transferring information into the computer and is the most basic purpose for which computers are used. Data entry operators are needed in almost every organisation using computers. People with good typing skills are best suited for these jobs.

What Careers Will I Qualify For With an Information Technology Degree?

"Information technology jobs lie within one of four main career paths. The broad Computer Science Application grouping includes systems analysts, computer programmers, computer scientists, software engineers and database/network administrators.

The remaining positions fall under either the Management/ Supervisory group (*e.g.*, project or technical managers), in Customer Service and Support (computer support specialists), or in Sales/ Relationship Management (sales engineers)."

Possible Job Titles for Associate's or Bachelor's IT Degree Holders / Entry Level Job Titles.

Here is a sampling of jobs you for which you may be qualified with a degree in Information Technology.

Use this for inspiration, remembering that this may represents some, but certainly not all, of the careers you can consider.

- Application Developer
- Business Analyst
- Computer Service Technician
- Database Analyst
- Database Developer
- Help Desk Analyst
- IT Business Analyst
- IT Specialist
- SQL Server DBA
- Junior .ASP Developer
- Junior .NET Developer
- Network Systems Analyst
- Project Manager
- Technical Analyst
- Software Tester
- Industrial Designer
- User Interface Designer
- Web Metrics Analyst
- Web-Internet Developer
- Wireless Support Analyst

Possible Job Titles for Advanced IT Degree Holders:

- Chief Information Officer
- Computer Engineer
- C++/UNIX Software Engineer
- Data Processing Manager
- Database Administrator

- Hardware Engineer
- Industrial Designer
- Information Architect
- Instructional Technology Manager
- .NET C# Developer
- Network Administrator
- Oracle Developer
- QA Engineer
- Robotics Engineer
- Senior Information Technology Engineer
- JAVA/J2EE Developer
- Software Developer
- Systems Consultant
- Systems Manager
- Telecommunications Engineer

These days, it seems as if everything is controlled by computers. Not in a bad way, of course: Stanley Kubrick's worst fears as witnessed by *2001: A Space Odyssey* won't be coming true anytime soon. But the fact remains that from the databases maintained by the IRS to the purchasing records of supermarkets to the information doctors keep of their patients, computers have become the single essential tie that binds the world together. It therefore should come as no surprise that a degree in information technology is more marketable now than it ever has been in the past.

Information technology "is the study, design, development, implementation and support of computer-based information systems to address real -world problems". In other words, it is the science (and, indeed, some might even say the art) of working with computers in order to facilitate the easier and more efficient use of them by non-professionals and professionals alike.

Niche Areas: As has been noted several times already, there are many specific areas of the IT field on which you may choose to focus. They include, but are not limited to, the following:

- Graphics and Imaging
- Operating Systems
- Personal Information Management
- Publishing

- Document Management
- Statistical Analysis & Mathematical Modelling
- Document Integration
- Administrative Systems

Over the course of the past decade, more and more colleges and universities have begun offering degree programmes in information technology. This is because of the ever-increasing importance of the skill-sets IT practitioners possess. An associate's degree in IT will prepare you for further study at the undergraduate level, as well as provide an adequate base of knowledge that you may attempt to parlay to a job as an entry-level programmer or database manager.

A bachelor's degree in IT affords you the opportunity to learn about all aspects of the field and to begin the process of narrowing down your areas of specific interest so that you can attain a certain amount of expertise. Graduate degrees in IT are generally either for those who wish to pursue collegiate teaching or research, or who wish to eventually work in the highest levels of the technology world—government computer development or computer and software design, for example.

Typical Admissions Requirements: Getting into an associate's degree programme in IT is not terribly difficult at all. As long as you have an interest in the field and a bit of previous experience with computers, then you should be fine. At the higher levels of education in the IT field, however, competition can become rather stiff.

Indeed, because of the high skill level of many applicants and the popularity of the field in general, there are likely to be many people with high levels of ability vying for a small number of spots. Don't let that discourage you, however. If you are truly interested in the field and even if you don't get in the first time you try, you should not throw in the towel. Perhaps with a bit of work experience and a bit of additional effort, you, too, will be on your way to a degree in information technology.

Careers in This Field: Because of the many and wide-ranging applications of computers and software, there is a nearly infinite variety of career options available to graduates of IT programmes. They include, but are not limited to, the following:

- Computer Software Engineer, Systems Software
- Computer Systems Analysts
- Computer Support Specialists

- Computer Programmers
- Computer and Information Systems Managers
- Computer Software Engineers, Applications
- Database Administrators
- Computer Security Specialists
- Computer and Information Scientists, Research

Salary Ranges in this Field: Because of the wide variety of careers available to those possessing degrees in IT, it is difficult to narrow down how much they stand to make. A government employee, for example, may very well make less money than an IT expert working with, say, a publicly-held financial institution on Wall Street. Also affecting the amount of money you stand to eventually make is the specific area of IT on which you choose to focus. Therefore, in order to more fully and accurately gauge what you'll eventually earn, it is best to do research on as specific an area as possible in the IT`field.

Future Outlook: Because the IT field is so large and wide-ranging, it is virtually impossible to discuss the job outlook in general terms. However, the field of computer and information systems managers is as good a place to begin as any and the outlook for it is likely indicative of larger trends in the field. According to the United States Bureau of Labour Statistics:

"Employment of computer and information systems managers is expected to grow faster than the average for all occupations through the year 2014. Technological advancements will boost the employment of computer-related workers; as a result, the demand for managers to direct these workers also will increase. In addition, job openings will result from the need to replace managers who retire or move into other occupations. Opportunities for obtaining a management position will be best for those with computer-related work experience; an MBA with technology as a core component, or a management information systems degree; and strong communication and administrative skills.

Despite the downturn in the technology sector in the early part of the decade, the outlook for computer and information systems managers remains strong. To remain competitive, firms will continue to install sophisticated computer networks and set up more complex Internet and intranet sites. Keeping a computer network running smoothly is essential to almost every organization. Firms will be more willing to hire managers who can accomplish that.

Similarly, the security of computer networks will continue to increase in importance as more business is conducted over the Internet.

The security of the Nation's entire electronic infrastructure has come under renewed scrutiny in light of recent threats. Organizations need to understand how their systems are vulnerable and how to protect their infrastructure and Internet sites from hackers, viruses and other acts of cyber-terrorism. The emergence of cybersecurity as a key issue facing most organizations should lead to strong growth for computer managers. Firms will increasingly hire cybersecurity experts to fill key leadership roles in their information technology departments because the integrity of their computing environments is of utmost concern. As a result, there will be a high demand for managers proficient in computer security issues.

With the explosive growth of electronic commerce and the capacity of the Internet to create new relationships with customers, the role of computer and information systems managers will continue to evolve. Persons in these jobs will become increasingly vital to their companies. The expansion of the wireless Internet will spur the need for computer and information systems managers with both business savvy and technical proficiency."

History of Information Technology

The term "information technology" came about in the 1970s. Its basic concept, however, can be traced back even further. Throughout the 20th century, an alliance between the military and various industries has existed in the development of electronics, computers and information theory. The military has historically driven such research by providing motivation and funding for innovation in the field of mechanization and computing. The first commercial computer was the UNIVAC I. It was designed by J. Presper Eckert and John Mauchly for the U.S. Census Bureau. The late 70s saw the rise of microcomputers, followed closely by IBM's personal computer in 1981. Since then, four generations of computers have evolved. Each generation represented a step that was characterized by hardware of decreased size and increased capabilities. The first generation used vacuum tubes, the second transistors and the third integrated circuits. The fourth (and current) generation uses more complex systems such as Very-large-scale integration or System-on-a-chip.

Information Technology Today

Today, the term Information Technology has ballooned to encompass many aspects of computing and technology and the term is more recognizable than ever before. The Information Technology umbrella can be quite large, covering many fields. IT professionals perform a variety of duties that range from installing applications to designing complex computer networks and information databases. A few of the duties that IT professionals perform may include:

- Data Management
- Computer Networking
- Database Systems Design
- Software design
- Management Information Systems
- Systems management

A more extensive list of related topics is provided below.

Information Technology Title

The Commonwealth of Pennsylvania has made a substantial investment in information technology resources. Many state agencies have installed a mainframe computer, midrange computer and a local area network. Our computer systems and programming languages are continuously being evaluated and updated. Also, information technology specialists receive training in the latest technological advances.

State agencies have developed several dynamic and innovative training programmes to introduce new employees to their systems. Classroom training, specialized workshops and vendor courses form the core of the training. Employees have the opportunity to follow a progressive career path that can lead to higher levels of computer programming or computer systems analysis and management.

Many of the latest concepts in the computer field are under development or nearing implementation including integrated communications networks, client server technology, distributed technology, imaging technology, CASE tools, computer assisted design, artificial intelligence and geographic information systems. Pennsylvania state government has also been recognized as a national leader in its' management of the Year 2000 date modification. In addition, the Commonwealth is committed to providing information to the residents of Pennsylvania by utilizing the technological opportunities available through the Internet.

Information technology (IT) is a label that has two meanings. In common usage, the term "information technology" is often used to refer to all of computing. As a name of an undergraduate degree programme, it refers to the preparation of students to meet the computer technology needs of business, government, healthcare, schools and other kinds of organizations.

IT professionals possess the right combination of knowledge and practical, hands-on expertise to take care of both an organization's information technology infrastructure and the people who use it. They assume responsibility for selecting hardware and software products appropriate for an organization. They integrate those products with organizational needs and infrastructure and install, customize and maintain those applications, so providing a secure and effective environment that supports the activities of the organization's computer users. In IT, programming often involves writing short programmes that typically connect existing components (scripting).

The shaded area extends down most of the right edge as it focuses on the application, deployment and configuration needs of organizations and people over a wide spectrum. Across this range (from organizational information systems, to application technologies and down to systems infrastructure), the role of the IT specialists has some overlap with IS, but IT people have a special focus on satisfying human needs that arise from computing technology. In addition, the IT shaded area goes leftwards from application towards theory and innovation, especially in the area of application technologies. This is because IT people often develop the web-enabled digital technologies that organizations use for a broad mix of informational purposes. This implies an appropriate conceptual foundation in relevant principles and theory.

Honeywell Technology Solutions Lab

Career: Information Technology in the Honeywell Technology Solutions Lab (HTSL) consists of developing enterprise middleware/ Internet technology solutions and applications for various Honeywell businesses worldwide. The group provides value by developing solutions that result in increased productivity—mainly in custom web application development, web interface, legacy applications, enterprise application integration and information portals.

Honeywell gives motivated employees increased responsibilities. We consciously identify and develop employees from within to grow into tomorrow's leaders and managers. As a result, the typical employee enjoys a career with exponential growth, fuelled by motivation, capabilities and a desire to excel.

Honeywell Technology Solutions Lab, Honeywell's key regional arm of excellence, has operations in Bangalore and Madurai, India; Shanghai and Beijing, China; Minneapolis, Minnesota and Phoenix, Arizona, USA; Singapore; and Brno in the Czech Republic. The Bangalore-based engineering and technology lab continuously enhances expertise in mission-critical technologies in the aerospace and controls domain. The engineering and technology lab in Bangalore has more than 3,000 employees with world-class facilities with state-of the-art hardware, software and communication infrastructure.

Career Focus: A career in Information Technology with a focus in Corporate Honeywell Technology Solutions Lab (HTSL) is an area integral to Honeywell. HTSL provides technology, product and business solutions while meeting global standards of quality, innovation and lifetime performance. Some of the applications include real-time process control, embedded real-time systems, system software and engineering tools. Research teams explore advanced technologies and frontiers in software to support Honeywell units worldwide and develop new technology-based products.

The engineering services initiative aims to provide best-in-class product and component analysis, design and prototyping in mechanical engineering and total engineering capability in electronics hardware.

HTSL is ISO 9001 TickIT certified and operates at a SEI-CMM Level 5 maturity with robust systems in place. It uses Six Sigma and Design for Six Sigma (DFSS) techniques to improve output in all areas. It recently earned PCMM Maturity Level 5 and was voted as one of the Top 10 Great Places to Work in a survey conducted by The Great Place to Work Institute, USA and Business in 2004.

Career Path: Career paths within the Honeywell Technology Solutions Lab vary greatly. Engineers with graduate, postgraduate, doctoral, or management educational qualifications join HTSL as individual contributors in the software programming, testing, research & development, or analytics business units. After gaining considerable experience and having a favourable performance rating, they can choose to go into a leadership role for technology, people, or process within their respective domains. From there, they can continue to grow. Successful leaders and experts can ultimately become strategic leaders and chief engineers.

Individuals may also move horizontally across roles. People who are currently in the role of leaders can also start afresh as individual contributors in a new technical area, domain, or function. One example is of a Mechanical Engineer at the Indian Institute of Technology. He began his career as a Software Development Engineer in the Home & Building Controls (H&BC) business for Automation & Control Solutions (ACS). Within two years, he started leading a team of development engineers and later began leading entire project teams. His overall technical expertise in business and people management skills allowed him to become the head of the H&BC business that culminated in leading the entire ACS business unit at HTSL.

Information Technology: As it is

Size and scope: The IT workforce is almost 1.2 million strong. The sector is divided into two parts:

- those working in the IT industry;
- IT professionals working in other areas.

Inside: There are almost 580,000 people working in companies in the UK whose primary function and business is IT; this accounts for approximately 2% of UK employment.

IT professionals typically fill roles in:

- IT services (*e.g.*, internet and web design services);
- technology development;
- systems analysis and testing;
- programming.

Just under a third of those working inside the industry are employed in other, less technical occupations, for example in:

- sales and marketing;
- consultancy;
- customer support;
- management.

Outside: There are almost 590,000 IT professionals working in other sectors in the UK. These are people whose primary role is IT orientated – they often work in IT departments or as IT support staff within organisations.

Outside the IT industry, the largest numbers of IT professionals can be found in manufacturing, financial services, retail and the public sector, although there are opportunities in just about every sector. Areas of work include:

- development (*i.e.*, creating systems, networks and applications);
- operations (*i.e.*, running and improving the speed of access to systems, networks and applications);
- user support.

Employment Trends: Four out of ten UK businesses employ IT professionals. Levels of employment in specific IT roles can be distributed as follows:

Table: *Employment levels in IT Q2 2005*

IT/Telecoms managers	32%
Technical support staff	28%
PC support staff	9%
Systems designers	7%
Systems developers	6%
Programmers	5%
Software engineers	5%
Operations staff	3%
Networking staff	3%
Internet professionals	2%
Database staff	1%

(*ICT Inquiry issue 3—Q2*, e-skills UK, 2005)

Location: London and the South East continue to dominate employment in IT with around 40% of IT professionals working in these two regions; Northern Ireland, Wales and Scotland account for 11% of IT employment in the UK. (*Labour Force Survey*, Office for National Statistics, 2004)

Graduate Recruitment: The number of IT jobs advertised in 2004 fell from the previous two years. However, there is substantial evidence to suggest this trend has begun to change for the period 2005-2006. Salaries compare well. The average graduate starting salary in 2005 was £28,095. Salaries were higher again for IT jobs in large organisations. (*TARGET Graduate Trends Survey 2005/6: IT sector*, GTI Specialist Publishers, 2006) IT professionals tend to be better qualified than the general workforce, with more than half having completed some form of higher education. According to what do graduates do? 2006, based on a national survey conducted six months after graduation, 40% of those who graduated with an IT-related degree in 2004 were working as IT professionals.

The most common roles for all new graduates are:

- software engineer;
- computer/IT consultant;
- computer programmer;
- systems analyst;
- computer analyst/programmer;
- computer operations manager.

Graduate Perceptions: Comments from graduates about what they enjoy most about working in IT reinforce perceptions of a dynamic industry that is often informal, but always focused on achievement. Highlights include:

- constant challenges and changes;
- enormous variety of work;
- relaxed work environment;
- culture of delivery and getting things done;
- part of a forever evolving and fast-moving industry;
- enthusiasm and intelligence of colleagues;
- good work/life balance.

This final point is significant as IT often has a 'long hours culture'. The cyclical nature of projects means that extra work may indeed be necessary to meet project deadlines or to provide 24/7 support; in some areas, notably the games industry, long hours are pretty much the norm. On the whole though, case studies show that many graduates work a fairly standard day.

Culture

Gender: Women form less than 20% (one in five) of the IT workforce, which is significantly low in comparison with the 45% of the UK workforce overall. Women do, however, have a higher presence in operations and user support technician roles. (*Quarterly Review of the ICT Labour Market Issue 14—Q3*, e-skills UK, 2005)

Training: The IT industry invests significantly more in training than the average invested by all other industries. This often includes training in new technology. The type of training offered will depend on the business priority of the organisation; there is tremendous diversity between companies in different areas of the sector. The following analysis of business priorities is from e-skills UK:

- IT services: customer satisfaction: revenue and profit; competitiveness; services development.
- Software product development: speed to market; product differentiation; cost control; long-term product profitability.
- IT sales and marketing: revenue; market share; win rate; customer retention.
- In-house IT: service levels and availability; cost effectiveness; systems strategy.

Skills: The industry tends to favour proven experience and ability over education and qualifications. In some roles, non-IT graduates compete with IT graduates based on the skills they have to offer, including technical and analytical skills and not necessarily degree subject studied. When recruiting for roles, almost 65% of employers seek graduates from any degree discipline. (*TARGET Graduate Trends Survey 2005/6: IT sector*, GTI Specialist Publishers, 2006)

Gaps: There is, at present, a significant IT skills gap in the UK. The British Computer Society states that current graduates do not have the technical skills the industry needs. Skills required by programmers and application developers, such as Java, are particularly lacking. This skills gap has resulted in the delay by some companies of the development of new products and services.

This is reinforced in *IT Insights: Trends and UK Skills Implications* (e-skills UK/Gartner Consulting, 2004). The report further suggests that the educational infrastructure needed to meet future skills requirements is not yet in place and the number of those applying and being accepted onto IT-related higher education courses is in decline.

Current Trends: Perhaps the most significant influences on the industry at the moment are:

- outsourcing—the concept of taking internal company functions and paying an outside firm to handle them;
- off shoring—the relocation of IT services to a lower cost location, usually overseas.

According to a recent survey by Capgemini, the UK currently outsources around 8% of its operations overseas, compared with the European average of 2%. This figure is growing annually. (*European CIO Survey: Views on Future IT Delivery*, Capgemini, 2006)

The UK has been slower and less aggressive in adopting and exploiting technology than other leading countries; businesses have failed to invest in IT, preferring to outsource instead. However, as the relationship between business productivity and company investment in IT becomes more apparent, there has been a drive towards greater IT innovation and investment. Business projects without IT as a major component are rare; companies are likely to spend more on new IT systems to inform their strategic planning and to implement business goals. The introduction of ID cards continues to be a current and problematic issue for the UK government; it is its biggest technological challenge to date, along with developing the extensive IT structures required to make the 2012 London Olympics a success.

Graduate Careers in Information Technology (IT)

The future for students wishing to pursue a graduate career in information technology provided they develop the skills in most demand in the workplace. According to a recent survey of 5,000 staff by the Association of Technology Staffing Companies, salaries for IT staff have risen an average of 15% over the last year due to increasing demand for senior project managers and business analysts. There has never been a better time to embark on a graduate career in IT as demand for management information systems, IT managers, business systems analysts and project managers is rising dramatically. A graduate career in IT will require you to use a variety of skills in addition to those acquired in your degree.

IT touches more areas of business than almost any other discipline. In today's business world many companies are interested in recruiting well-rounded staff who have business focused skills and can demonstrate an understanding of how IT can benefit the business as a whole and an understanding of how implementation of IT systems will impact on the organisation. People skills, self management and an ability to see IT as a means to an end, rather than an end in itself are all equally important as the traditional core IT skills.

There are currently several routes to a career in the UK IT industry including full time study, part-time study whilst in employment and on the job training. The following are just some of the options open to those interested in an IT career:

Graduation with a non-IT related subject.

An accredited degree.

Training in a specific IT skill which is in demand.

Transfer to an IT department within a company.

A year in industry/gap year experience prior to University.

Many companies are happy to recruit graduates with non IT related subjects who are willing to undertake additional IT specific training through postgraduate or professional training courses such as those offered by Information Systems Examination Board (ISEB)–a BCS subsidiary, which offers qualifications in systems analysis and design and project management, among others. It is also worth considering undertaking a language; the IT profession is global and having additional language skills can be useful in gaining employment.

For students interested in a "client facing role" a more business orientated degree such as one in finance, management or another engineering subject may prove a better grounding than a pure IT degree, providing students with a good insight into business and a better understanding of the sort of problems their future client's are likely to face. Whichever option taken, the key to improving the chances of being recruited is to achieve the best degree possible and take advantage of any work experience offered as part of the course or during the vacations.

The most direct route into the more traditional "hard" IT roles such as software development, infrastructure and research posts, is a degree in computing followed by an application to a graduate recruitment programme. An accredited degree offers a core of studies seen as the minimum necessary for the foundation of a professional career in the industry, together with specialist content in one or more areas studied in-depth; it will include an appropriate mix of engineering principles, design and problem-solving and practical work. BCS accredits honours degrees in the IT area for Chartered Engineer (CEng) status and professional membership. BCS also accredits degrees for Chartered Scientist. In addition, there are four-year MEng courses.

Many degree courses include a year spent in industrial training in a company with a computing environment prior to the final year. Some employers will offer sponsorship for the final year after a successful placement, with or without a guarantee of employment on graduating. After study, there are opportunities to join a graduate development programmes offered by companies for new graduates aiming to work in IT. Alternatively, graduates with a good class degree may consider studying for a PhD or moving into industrial research.

A specific training course in a skill which is in demand offers a quick way to enter the industry. Internet website design and enterprise resource planning (ERP) are just two examples of skills currently in high demand, although the market is continually changing. However, as demand for particular skills can be short-term, it is advisable to continue developing additional skills and gain broad experience.

Organizations such as Microsoft (for Windows), Novell (for networking) and Oracle (for databases) offer vendor certification courses. The Information Systems Examination Board (ISEB) offers qualifications in systems analysis and design and project management and learning can be done via classes, computer-based training or distance learning. Courses like these which focus on specific computer skills are aimed at the experienced practitioner and are not suitable for the beginner. Finally the European computer driving licence (ECDL) is a users' qualification recognised across the world and offers a starting point and evidence that students are prepared to train and build skills.

For information relating to skills required for each specific role in graduate careers in IT, the Skills Framework for the Information Age (SFIA), the high level UK Government backed competency framework, describes the roles within IT and the skills needed to fulfil them.

SFIAplus contains the SFIA framework of IT skills plus detailed training and development resources to provide the most established and widely adopted IT skills, training and development model reflecting current industry needs. If you are interested in pursuing a graduate career in IT, then don't delay, start enhancing your knowledge and skills with additional courses and qualifications to give your career the best possible start.

Technology, Telecoms and Contact Centres

Over the last few years, these industries have been among the fastest growing and developing sectors in the country. They have an impact on all industries.

- About two per cent of all Scottish jobs are in the IT and telecoms industries. (48,700)
- There are also IT professionals employed in other sectors; over half a million in the UK.
- The UK is the third largest telecoms market in the world.
- There are 56,000 contact centre jobs in Scotland; almost 86 per cent of these are in the central belt of Scotland.
- Most industries have contact centres; financial services, with 30 per cent of the total number of jobs, remains the most important user of contact centres.

Areas Included: The IT and telecommunications sector is responsible for:

- the installation and maintenance of communication networks
- the installation and support for new computer software and hardware

These industries have changed the way we lead our lives and altered the way in which business is conducted; you can play videos on mobile phones and hold face to face meetings with business colleagues in other countries.

There is almost no aspect of our lives that is not affected by this industry.

In Contact Centres, staff no longer just use a telephone; they also use the Internet, e-mail and SMS messaging.

- More than half of all jobs are both professionals and associate professionals.
- The most common IT professional job role in the UK is "software engineer".
- Contact Centre staff require high levels of technical, language and personal skills.

For a full list of jobs in this industry see below.

What's it like to work in the information technology, telecommunications & contact centres industry

- This industry will appeal most to those with an interest in technical work.
- Most jobs are full time – 86%.
- Most employees are men – 62%.
- Self employment in IT is slightly higher than the average.
- The average age of a worker is 36 – younger than the average of 40 for Scotland as a whole.
- In contact centres, three fifths of the workforce is female (61%).
- The prospects for graduates are good. In the last two to three years, one in four employers had recruited a graduate.
- Employers are less likely to recruit school leavers. Less than one in ten employers have recruited a school leaver in the last two to three years.

Future Trends

- Employment is expected to increase by more than 20% in the next ten years.
- Future skills shortages are expected to be in systems integration, networking and business analysis.
- Despite some call centres moving overseas and changes in technology; the industry is predicted to continue to grow.

Information Technology Jobs

At Argonne, we are building a cutting-edge computing environment enabling scientific breakthroughs and supporting laboratory business. We are looking for top-quality, highly motivated people who seek personal and professional fulfilment in their careers. Your problem-solving and leadership skills as well as a commitment to continuous improvement and customer satisfaction are needed.

Argonne is known throughout the world for leading-edge science and technology research. Argonne's future depends on high-caliber people who contribute to every aspect of the organization. Argonne is a challenging and rewarding career choice, a workplace of which we can all be proud.

Current Openings

Web Developer: As a member of the Computing and Information Systems Division, this position develops and supports web-based applications for Argonne's business systems and public web pages. This person will work with users across the Laboratory to analyze their requirements, define projects and implement or customize web-based applications.

Linux System Administrator: As a member of the Computing and Information Systems Division, this position implements, enhances and maintains robust, redundant, Unix-based web infrastructure and services. This person develops, implements and supports web-enabled tools for distributed Unix systems administration and participates in the conceptualization, planning and implementation of Laboratory-wide distributed UNIX computing infrastructure.

Senior Software Developer: As a member of the Leadership Computing Facility (LCF), this position builds computational tools and develops methods for the solution of computational science problems on parallel and distributed computers.

Senior Systems Administrator: As a member of the Mathematics and Computer Science Division's High-Performance Computing (HPC) team, this position is dedicated to the technical operation, support and development of the Transportation Research and Analysis Computer Center (TRACC). Special emphasis is given to leading efforts in building and maintaining systems and clusters, creating technical documentation for end-users and internal use and providing technical support to the users of HPC resources. This position is stationed at Dupage Technology Park.

Senior Software Developer: The Mathematics and Computer Science Division has two openings for senior software developers who will work with members of the Radix Laboratory, which develops software for petascale systems. They will be responsible for designing, implementing, debugging and testing new components.

Candidates should have considerable experience with Linux, computer science, software engineering and the C programming language. Familiarity with parallel programming, high-performance computing, operating systems, high-performance networking and MPI is highly desired. The appointees will work as part of an integrated, multi-institution research team whose focus extends from the investigation of algorithms to parallel scientific simulations.

Computational Scientist: As a member of the Leadership Computing Facility (LCF), this position assists and collaborates with Leadership Science Project Teams to enable breakthrough science and engineering research.

You will need comprehensive knowledge of high-performance computational science in one or more of these fields: computational materials science, computational biology, computational lattice quantum chromo dynamics and/or computational astrophysics. Also required are considerable skills in verbal and written communications, collaboration, leadership in interdisciplinary research and the solution of computational science problems using scalable parallel computers and large data systems.

Manager, Facility Operations and Networking: As a member of the Leadership Computing Facility (LCF), this position is responsible for operation of the LCF computing, storage and network hardware systems. This position requires comprehensive skills in leading system support efforts; deploying and operating advanced computing systems; and developing effective operations processes. Also required is comprehensive experience in supercomputing systems administration, including configuration, installation, monitoring and load optimization, leading to a robust and stable computing environment.

Manager, Systems Software and Integration: As a member of the Leadership Computing Facility (LCF), this position leads the team responsible for the operation of the high-performance data storage disk and tape systems and the integration of new experimental systems software on the production systems. Comprehensive knowledge of high-performance systems software and data storage systems is required, along with demonstrated organizational and communications skills and flexibility in coordinating a spectrum of activities. Experience in national facility team leadership is highly desired

Content Management Administrator: The Computing and Information Systems Division is seeking a Content Management Administrator to be part of the newly formed Web Services team to manage, customize and support the Stellent content management system.

The Web Services team is responsible for designing, creating and managing web systems and content management systems that support the Laboratory. The ideal candidate will have strong technical skills in web technologies and collaboration tools and content management systems. Considerable experience with web development—Java, JSP, Scripting Languages, Portlets, Web Services, HTML, XML.

Optical Network Design Specialist: The Computing and Information Systems Division is seeking an optical network design specialist who will be responsible for planning, implementing, operating and testing complex optical transmission systems.

Work will include advanced fiber-optic, digital and optical cross-connect and network management systems as well as transport customer interconnect facilities. Responsibilities include system architecture and design, engineering specifications, installation and as-built documentation and verification and system monitoring methodologies.

The ideal candidate will be able to work independently with limited guidance and balance day-to-day operational needs with long-term project goals.

Jobs in Information Technology: If you are looking for some information about a specific type of job or are just curious about other career paths, review the following positions to help choose the right one for you.

Business Teachers, Postsecondary: Teach courses in business administration and management, such as accounting, finance, human resources, labour relations, marketing and operations research.

Computer and Information Scientists, Research: Conduct research into fundamental computer and information science as theorists, designers, or inventors. Solve or develop solutions to problems in the field of computer hardware and software.

Computer Hardware Engineers: Research, design, develop and test computer or computer-related equipment for commercial, industrial, military, or scientific use. May supervise the manufacturing and installation of computer or computer-related equipment and components.

Computer Operators: Monitor and control electronic computer and peripheral electronic data processing equipment to process business, scientific, engineering and other data according to operating instructions. May enter commands at a computer terminal and set controls on computer and peripheral devices. Monitor and respond to operating and error messages.

Computer Programmers: Convert project specifications and statements of problems and procedures to detailed logical flow charts for coding into computer language. Develop and write computer programmes to store, locate and retrieve specific documents, data and information. May programme web sites.

Computer Science Teachers, Post-secondary: Teach courses in computer science. May specialize in a field of computer science, such as the design and function of computers or operations and research analysis.

Computer Software Engineers, Applications: Develop, create and modify general computer applications software or specialized utility programmes. Analyze user needs and develop software solutions. Design software or customize software for client use with the aim of optimizing operational efficiency. May analyze and design databases within an application area, working individually or coordinating database development as part of a team.

Computer Software Engineers, Systems Software: Research, design, develop and test operating systems-level software, compilers and network distribution software for medical, industrial, military, communications, aerospace, business, scientific and general computing applications. Set operational specifications and formulate and analyze software requirements. Apply principles and techniques of computer science, engineering and mathematical analysis.

Computer Support Specialists: Provide technical assistance to computer system users. Answer questions or resolve computer problems for clients in person, via telephone or remote location. May provide assistance concerning the use of computer hardware and software, including printing, installation, word processing, electronic mail and operating systems.

Computer Systems Analysts: Analyze science, engineering, business and all other data processing problems for application to electronic data processing systems. Analyze user requirements, procedures and problems to automate or improve existing systems and review computer system capabilities, work-flow and scheduling limitations. May analyze or recommend commercially available software. May supervise computer programmers.

Information Technology Manager

Information technology managers make sure computer departments run smoothly and efficiently. They may work with systems analysts to improve computer systems. They also manage databases, organise staff training, manage budgets, arrange computer maintenance and put into place backup systems in case an IT fault develops.

Work Activities: Information technology (IT) managers make sure that computer departments run smoothly and efficiently. They have overall responsibility for the use of IT within the company.

The IT manager first has to make sure that the company has all the right equipment it needs to be as efficient as possible. IT managers therefore need a very broad knowledge of different IT systems; they must keep up-to-date with advances in the technology. They are also likely to be in charge of a budget, spending money wisely to bring the most appropriate technology into the company. They work closely with equipment suppliers, negotiating the sale and any aftersales services, like technical support in case there are any faults with the equipment.

As well as buying new systems, IT managers keep a close watch on the technology the company already has. They think about the company's needs and identify areas where new technology could support people's work. They may ask a systems analyst to visit the company to do an in-depth study of the existing technology and come up with suggestions to improve the situation.

Information technology managers work as closely with people as they do with machines. They make sure people are properly trained and supported in their use of IT; they may ask a computer trainer to visit the company to teach people how to use a specific system or software product. Managers are responsible for setting quality standards and for making sure people complete their work within deadlines and budget limitations.

They are also responsible for the accuracy and security of data within the organisation. A strict data protection law controls the use and security of information held on databases; it's up to the manager to make sure only authorised people can look at the data. Also, members of the public have the right to access any information about them on a company's database, so managers may have to negotiate this access with them. IT managers must be able to cope quickly and efficiently if there are any problems with the company's computer systems. They must set up backup systems to make sure no data is lost if there is a fault.

Personal Qualities and Skills: To be an information technology manager, you must have a broad knowledge of computer systems and software products. Just as importantly, you must be willing to keep up-to-date with developments in IT. You will need strong communication skills, to negotiate with equipment suppliers and to work closely with people throughout the organisation. You must be able to explain things clearly and concisely to people who may have little knowledge of computers and be able to ask the right questions to assess their training needs. Information technology managers need very good organisation skills to plan work, arrange meetings with other professionals (like systems analysts or computer trainers) and set deadlines and targets.

You must be able to cope well under pressure, for example, if the system develops a fault.

Pay and Opportunities: The pay rates given below are approximate. IT Managers earn in the range of £24,500—£30,500 a year, rising to £41,000—£55,500. Higher earners can make around £73,500 a year. Salaries may include performance related pay, profit sharing or company bonuses.

IT managers usually work 35—37 hours Monday to Friday, though occasional late finishes may be required. Jobs exist throughout the UK, with employers in industry and commerce, including banks, building societies and insurance companies and in the public sector with local and central government departments, the NHS and public utilities. Consultancy and fixed-term contract work can be available for experienced managers.

The following information is sourced from government statistics. The government's definition of types of job is slightly different to that used in this programme. For this type of job, the most relevant Government-defined occupation is 'Information and communication technology managers'. In 2001 there were 9,400 people working as Information and communication technology managers in Scotland. 43 in every 10,000 employees working in Scotland were employed as Information and communication technology managers. Estimates of projected future job openings requiring new entrants to the jobs market are available, but not at the level of detail of 'Information and communication technology managers'. It is estimated that over the five years between 2003 and 2008 there will be a need for 21,000 new employees to fill vacancies in the broader occupational type of 'Functional Managers'.

1 out of every 20 employees works part time in this type of job.

Entry Routes and Training: Most information technology managers are graduates, although few enter this job straight after graduation. Most people first gain experience in other IT careers, especially systems analysis (because this job involves working closely with people and using management skills).

Many managers are members of the Institute for the Management of Information Systems (IMIS). IMIS offers training leading to diploma, higher diploma and graduate diploma qualifications. For entry to the graduate diploma, you should have the IMIS higher diploma, a Higher National Diploma or Higher National Certificate in Computing (though you will need to support the HND or HNC with two years' relevant work experience), or a degree in any subject plus at least one years' relevant experience. The graduate diploma covers Human and Computer Interaction, Database Design, Information Systems Engineering and Business, Systems and Strategy.

Qualifications: For entry to a degree course the minimum requirement is 3 Highers (A-C) plus Standard Grades (1-3) in 2 other subjects. Entry requirements vary between courses and alternative qualifications may be accepted—check prospectuses for details.

Information technology managers have usually worked for some time in the computing industry, for example in systems analysis. According to government statistics, people working in this job have the following qualifications. Higher Education: 6 out of 10 employees have their highest level of qualifications attained at degree level or equivalent or higher.

Post-16 or Further Education: 2 out of 10 employees have their highest qualification at a level below a degree or higher degree or equivalent, but gained after the age of 16. School leaving age (16) qualifications or qualifications at SVQ3: 3 out of 10 employees gained their highest qualification at the age of 16 or have gained another qualification lower than a VQ level 3 since they were 16.

Adult Opportunities: There is no formal upper age limit for entry into this occupation.

Some employers will consider applicants with substantial relevant experience, even if they lack the usual academic or professional qualifications for entry.

Relevant experience can be as a systems designer/programmer, engineer or analyst, with a focus on one or more specialised areas of IT, such as networks or databases.

Experience as a team leader can be an advantage for entry into management level posts. For senior posts, taking the MBA (Master of Business Administration) can be an advantage, usually after a minimum of two years' IT management experience.

A range of manufacturer-accredited courses are available on an intensive basis, often flexible and part-time, including evenings and weekends. Distance learning is also offered by Abacus Learning Systems, with courses in IT management training, plus the British

Computer Society Professional Examination Certificate, Diploma and Graduate Diploma. Suitable applicants aged 21 or over may do a college or university Access course. No formal qualifications are required to enter an Access course, but you should check individual course details. They lead to relevant degree/HND courses.

Education, Training and Degrees

Individuals looking for employment as information technology specialists generally must have at least a bachelor's degree in computer science, computer engineering, or a similar field.

Certification in a specific technology is also helpful. Many companies, such as Microsoft, MySQL, Oracle, Cisco and Red Hat offer training and exams to help aspiring IT candidates become certified in their particular tools and technologies.

Individuals seeking a career in IT management should consider obtaining a master's degree in information systems or information technology, or an MBA with a focus on technology.

Explore Career Opportunities

The ongoing demand for online products and services, plus the emergence of new technologies, means that career opportunities in the field of information technology will continue to grow in the foreseeable future.

Areas of particular interest and growth include IT security, the Linux market, open source technologies, wireless and mobile devices, wireless networking and telephony.

A number of career opportunities exist for those interested in the field of information technology. From entry-level jobs in technical support or the help desk, to mid-level jobs such as network administrator, computer programmer and Web developer, to the higher-level jobs such as database architect or information security specialist, numerous positions exist within the IT domain. Below are a few of the most popular professions in this field.

Information Technology Specialist

Information technology specialists set up computer systems and networks, test new solutions, troubleshoot problems that users may have, purchase new hardware and software and install and maintain security systems, among other responsibilities.

They work with many people throughout a company, moving from computer to computer to find solutions to various computer problems.

Information Technology Manager

Information technology managers put both their technical and managerial skills to use by planning information strategies for a company, determining and working with budgets, coordinating projects and managing others.

Train for a new career in under two years!

LCCC offers many certificate training programmes that help you learn new career skills in a hurry. The short-term training opportunities are perfect for those who want to train for a good job, but aren't able to commit to long-term education or training. Credits earned for many Certificate of Proficiency or Certificate of Completion programmes can be applied to a related LCCC associate's degree or an LCCC University Partnership bachelor's degree.

Information Technology Training & Certifications

You can earn certificates, diplomas, bachelor's degrees, master's degrees and doctoral degrees in almost all of the programmes within the information technology field. Programmes in which you may obtain certifications or degrees include computer security, network administration, computer support, systems developing and analyzing, computer programming, telecommunications, computer science and web design. Individuals with advanced information technology training and certifications should enjoy highly favourable employment prospects. Employers tend to seek information technology specialists who can combine strong technical skills with good interpersonal and business skills. In some instances, if you have the right experience along with information technology training and certification, you can work in many of the computer occupations regardless of your level of formal education.

Job Outlook: The information technology field will be one of the fastest growing occupations over the next few years. The demand for specialists will increase due to technology enhancements and the integration of new technologies. Businesses and other organizations will need workers with information technology training to not only implement more sophisticated and complex technology into their offices, but also to maintain the networks and teach others how to operate them.

IT and Certification Programmes

Information Technology Training: Search this directory for information about Information Technology Training and find the perfect campus or online degree programme. Please request free information from as many schools as necessary to make the right decision for you—it is risk free and there is no obligation.

Information technology training prepares you for an exciting profession that involves hardware, software, services and supporting infrastructure to manage and deliver information. Information technology includes all computers, voice, video and data networks and the equipment and staff needed to operate them. It also includes all costs associated with operating and providing information technology, as well as developing, purchasing, licensing and maintaining software. Some examples of information technology commonly used includes telephone and radio equipment, software and support for office automation systems and the computers that run them, server hardware and software and computers and network systems used for educational purposes.

Master Certificate & Certificate Programmes

Our IT programmes are designed to help the individual gain the skills and industry certification that will put them in demand by employers. Key industry certifications include CompTIA A+, Network+ and Linux+ and Microsoft MCP, MCSA, MCSE and MCDST. Programmes vary in length and depth of training. Your Admissions Representative will help you select a programme that best matches your current skills and career goals.

Oracle Database Administration Master Certificate Programme: Prepare for a rewarding career in managing information-one of businesses most valuable assets. As an Oracle Database Administrator you'll play a critical role managing and administering databases, setting up and implementing backup and recovery strategies and establishing and maintaining appropriate security systems. In our Oracle Database Administration programme, you'll learn how to perform the major tasks needed to administer and support these database systems.

Understand Windows NT networking architectures and structures, PL/SQL (the language of Oracle) database administrative tasks, strategies for backing up an Oracle database, crash recovery procedures, security measures and ways of improving database performance through hands-on exercises. The Oracle Database Administration Programme uses the authorized Oracle curriculum. This programme is a ***Passport*** Programme-ask your counsellor for details.

Enterprise Solution Developer Master Certificate Programme: The Enterprise Solution Developer programme is targeted at the IT professional who wishes to upgrade their skills as a software developer to a true solution developer. Students will explore the Microsoft .Net framework with concentration on C# as well as the J2EE environment.

This intense programme educates the student on the architectural approaches used in enterprise projects and the technical skills needed to build, support and integrate enterprise applications. The programme focuses on technologies such as C#, JAVA, ASP.NET, ADO.NET, SQL Server, Oracle, XML and Web Services. Students who complete the programme have a sound, practical and hands-on experience as Enterprise Developers ready to fit into and succeed in today's complex information systems environment.

PC & Networking Support (CompTIA A+, Network+, Linux+ and Microsoft MCP Certifications) Master Certificate Programme: Get the skills needed to install and configure PC hardware such as disks, memory and network adapters and operating systems such as Windows and DOS. Learn how to install Microsoft Office, set up printing and troubleshoot common problems relating to Microsoft Office and e-mail. You'll also understand the fundamentals of computer networking, including LANs and WANs, TCP/IP and network administration. With this programme, you'll prepare for examinations, which lead to the CompTIA A+, Network+ and Linux+ Certifications, industry-wide vendor neutral, developed and sponsored by the Computing Technology Industry Association (CompTIA), as well as the entry-level Microsoft Certified Professional (MCP) certification.

PC & Networking Administration (CompTIA A+, Network+, Linux+ and Microsoft MCP, MCSA Certifications) Master Certificate Programme: Building on the PC & Networking Support classes, receive the career skills necessary to implement and administer Microsoft Windows 2003 Networking and Operating Systems Infrastructure. After successfully completing this programme you may pursue your Microsoft Certified System Administrator (MCSA) Certification.

PC & Networking Design (CompTIA A+, Network+, Linux+ and Microsoft MCP, MCSA, MCSE, MCDST Certifications) Master Certificate Programme: Building on the PC & Networking Administration classes, this programme trains you with the skills necessary to design and deploy Microsoft Windows 2003 Networking and Operating systems. After successfully completing this programme, you may pursue the most advanced certification from Microsoft—The Microsoft Certified System Engineer (MCSE) certification.

PC & Networking Design w/ Security (CompTIA A+, Network+, Linux+, Security +, MCP, MCSA, MCSE Certifications) Master Certificate Programme: A recent U.S. News & World Report article stated that Network Security is a "Profession in which jobs are projected to be plentiful for years to come." As one of the nation's top 8 career tracks, Network Security is a career with a future. Get the skills necessary to work in Network Security with this programme. You'll learn to design and analyze Network Security systems. After successful completion of this programme, you may pursue the Security Certified Network Professional (Security+) certification. This programme is a ***Passport*** Programme -ask your counsellor for details.

Careers, Jobs and Employment

Information Technology (IT) is a broad term that includes all aspects of managing and processing information and related technologies. Specifically, information technology professionals are responsible for designing, developing, supporting and managing computer hardware, computer software and information networks, such as the Internet. The realworld applications of information technologies can be found everywhere. In fact, IT is likely already a part of your life in ways you may not even be aware. Examples include computer software used to manage basic computer applications, computer generated animation in popular movies, networks and programmes that allow you to purchase online and satellites and systems that enable NASA to preform remote space exploration. There are a wide variety of career opportunities available for capable and experience IT professionals. Select a category below to review detailed job descriptions, career reviews, to identify useful technology degrees and to see employment forecasts.

How Outsourcing Affects Your Career in IT

In the United States, corporations plan to **outsource** many thousands of **Information Technology (IT)** jobs to outside firms. Most of these jobs will belong to so-called **offshore** organizations in India or Southeast Asia. The media buzz around IT offshoring and outsourcing continues to grow and take on a colder, more fatalistic tone as time passes.

As a current Information Technology professional in the U.S., or a student considering a future career in IT, outsourcing is a business trend you must fully understand. Don't expect the trend to reverse any time in the forseeable future, but don't feel powerless to cope with the changes either.

Changes Coming with Information Technology Outsourcing

Five or ten years ago, workers were attracted to the Information Technology field given the

1. challenging and rewarding work
2. good pay
3. numerous opportunities, the promise of future growth and long term job stability

Outsourcing will alter and is already altering each of these IT career fundamentals:

1. The nature of the work will change dramatically with offshoring; the future may be equally rewarding, or it may prove wholly undesirable depending on one's individual interests and goals.
2. Information Technology salaries will increase in the countries that receive outsourcing contracts and will decrease in the U.S.
3. Likewise, the total number of IT jobs will increase in some countries and decrease in the U.S, most future growth will happen outside the U.S and job stability will remain unclear everywhere until the outsourcing business models mature.

How to Cope with Information Technology Outsourcing?

IT workers in the U.S. may already be witnessing some impacts of IT outsourcing, but the future impacts are likely to be much greater. What can you do to prepare? Consider the following ideas.

- *Don't Panic:* The prospect of job searches or career changes can be quite stressful to Information Technology workers. IT students may understandably begin to question their choice of career. However, the more stress and worry a person takes on, paradoxically the more difficult it becomes to successfully reach their career goals.
- *Don't Wait for The Upturn:* So-called "experts" have been predicting a sharp upturn in the U.S. economy for several years. It's more likely that the sought-after economic recovery has already happened and we should expect to operate in the current climate for the foreseeable future.
- *Become a Generalist:* Years ago in Information Technology, specialization was king. Those with the heaviest technical backgrounds and loftiest job titles, like Enterprise Architects, commanded the highest salaries. Nowadays, a person is much better positioned if they are skilled in multiple areas of both technology and the business side of IT. Flexibility is now king.

- *Look to Smaller Organizations:* Fortune 500 companies will embark on the vast majority of outsourcing and offshoring deals. Outsourcing creates a substantial amount of overhead before the gains kick in and small companies can't afford to pay that price for the foreseeable future.
- *Start Your Own Business:* Uncertain economic times and times of big industry change are often the best ones for starting a new business, due to lower prices for capital, less competition and the natural emergence of big new market opportunities. All it takes is an entrepenurial attitude and a few good ideas.

Above all, whatever your chosen career path, strive to find happiness in your work. Don't fear the ongoing change in Information Technology just because others are afraid. Control your own destiny.

Starting or Building a Career in Computer Networking

Many view computer networking as one of the best and "hottest" career fields available today. Some claim that a serious shortage of qualified people to fill these networking jobs exists and these claims may lure some people into the fray hoping for an easy position with a fast-growing company. Don't be fooled! Debates over the actual extent of any "shortages" aside, networking involves mostly hard work and competition for the high-quality positions will always be strong. Continue reading to learn more about beginning or expanding a career in networking and pick up some valuable job-hunting tips that also apply to many other types of technical careers.

Job Titles: Several types of positions exist in networking, each with different average salaries and long-term potential and one should possess a clear understanding of these. Unfortunately, job titles in networking and in Information Technology (IT) generally, often lead to confusion among beginners and experienced folks alike.

Bland, vague or overly bombastic titles often fail to describe the actual work assignments of a person in this field.

The basic job titles one sees for computer networking and networking-related positions include

- Network Administrator
- Network (Systems) Engineer
- Network (Service) Technician
- Network Programmer/Analyst
- Network/Information Systems Manager

The Network Administrator: In general, network administrators configure and manage LANs and sometimes WANs. The job descriptions for administrators can be detailed and sometimes downright intimidating! Consider the following description that, although fictitious, represents a fairly typical posting:

Network Administrator—Hobo Computing

"Candidate will be responsible for analysis, installation and configuration of company networks. Daily activities include monitoring network performance, troubleshooting problems and maintaining network security. Other activities include assisting customers with operating systems and network adapters, configuring routers, switches and firewalls and evaluating third-party tools." Needless to say, a person early in their career often lacks experience in a majority of these categories. Most employers do not expect candidates to possess in-depth knowledge of all areas listed in the job posting, though, so a person should remain undeterred by the long, sweeping job descriptions they will inevitably encounter.

Comparing Roles and Responsibilities: The job function of a Network Engineer differs little from that of a Network Administrator. Company A may use one title while Company B uses the other to refer to essentially the same position. Some companies even use the two titles interchangeably. Firms making a distinction between the two often stipulate that administrators focus on the day-to-day management of networks, whereas network engineers focus primarily on system upgrades, evaluating vendor products, security testing and so on.

A Network Technician tends to focus more on the setup, troubleshooting and repair of specific hardware and software products. Service Technicians in particular often must travel to remote customer sites to perform "field" upgrades and support. Again, though, some firms blur the line between technicians and engineers or administrators. Network Programmer/Analysts generally write software programmes or scripts that aid in network analysis, such as diagnostics or monitoring utilities. They also specialize in evaluating third-party products and integrating new software technologies into an existing network environment or to build a new environment. Managers supervise the work of adminstrators, engineers, technicians and/or programmers. Network/Information Systems Managers also focus on longer-range planning and strategy considerations.

Salaries for networking positions depend on many factors such as the hiring organization, local market conditions, a person's experience and skill level and so on.

High School and College Education: Those interested in networking careers can benefit greatly from earning a college degree. Most university programmes don't offer a degree in Computer Networking *per se* and the precise name of the degree varies significantly from institution to institution. Four-year degree programmes suitable for the computer networking field usually involve a variation on one of the following:

- Computer Science
- Electrical and Computer Engineering
- Information Systems
- Communications Science
- Telecommunications, Telecommunications Management
- Telecomputing

As an alternative to a general four-year degree (that covers a variety of technical subjects besides computer networking), some institutions offer shorter-term programmes focused specifically on networking topics. Until recently, computer networking courses were only found in post-secondary education. Nowadays, though, high school students have the opportunity to take networking courses too. These classes can be quite substantial, involving among other things configuring routers and switches, installing wire, network diagnostics, monitoring network activity and working with various network protocols and operating systems.

Which Programme Is Best?: Is a college degree worth the investment, or is a shorter, more focused curriculum the way to go? Opinions vary. A four-degree can demonstrate to prospective employers a level of dedication and long-term flexibility that a short programme cannot. On the other hand, a more focused programme can teach the basic networking skills quickly and allow more time for on-the-job experience.

Certifications: Network administrators and managers in particular have grown fond of networking-based certifications like Microsoft MCSE and Cisco CCNA. In general, to gain and keep a certification one must pass a lengthy (usually multiple-choice question) paper exam, then pass recertification exams at periodic intervals (usually every two or three years). A person has the choice of preparing for the exam through self-study or by enrolling in a certification course or "programme" run by a training organization (sometimes within high-tech companies themselves). Taking any certification exam involves paying a test "sitting" fee (usually in the range of $100 to $300 USD) and employers sometimes reimburse their employees for this cost.

Certifications are designed to accredit someone for a certain amount of industry experience that they've already gained. Some of the programmes will even make recommendations to this effect, typically one to two years of prior background for the entry-level certifications. However, experience is not strictly required. Some have criticized the entry-level exams for being too "bookish" in this respect, too easy to pass without prior hands-on experience.

Which certification is best? MCSE? CCNA? Something else? Again, the answer depends on the individual's interests and also the preferences of hiring companies. Some ambitious students of networking avoid this problem by acquiring multiple certifications... sometimes as many as five or more! Be aware, though, that certifications are an incomplete substitute for formal education and industry experience. Ideally, one will acquire a few certifications as part of a balanced overall mix of education and career experience.

Many companies, particularly larger ones, offer their employees ongoing training opportunities. The employer will either build their own courses or will bring in an outside company to hold the training. These courses are typically focused on a specific product technology or tool, or on the specific technical information needed to pass a certification exam. One could argue it is preferable for the beginning networker to focus on general technologies at first rather than certifications, as companies in these case likely prefer to train employees "their own way" anyhow.

Networking Experience: The common lament of job seekers, that "employers only hire people with experience, yet the only way to gain experience is to get hired" applies in the computer networking field as well. Despite optimistic statements that one hears frequently regarding the number of available jobs in IT, landing an entry-level position can still prove difficult and frustrating.

One way to gain networking experience is to pursue a full-time programming or help desk "internship" during the summer months, or a part-time "work study" job at school. An internship may not pay well initially, the work may turn out to be relatively uninteresting and it is very likely one will not be able to finish any substantial project during the limited time there. However, the most important factor to consider is the training and hands-on experience such a job offers. The mere fact a person invests their time in this way, demonstrates the dedication and interest employers like to see.

The better the position, the more likely multiple candidates will apply for it, even if the job entails only part-time work. A good way to "stand out" from the competition is to demonstrate prior work and accomplishments, even if these involve projects done on one's own time. A person can start with a class project, for example and extend it in some way. Or they can create their own personal projects, experimenting with networking administration tools and scripts, for example.

Explaining Experience: One of the most overlooked skills in computer networking is the ability to explain technical information. Whether verbally, through e-mail, or in formal writing, networkers that communicate well gain a significant advantage in building their careers. For the beginning networker, the most obvious benefit of good communications skills is realized in job interviews. Being able to talk with people about technical subjects can be hard to do, but as one gains skill in answering impromptu questions, one builds confidence and relaxes, making one that much better prepared for career advancement. It is a good idea to periodically engage in job interviews for this reason, even if the position involved does not seem particularly appealing. Likewise one should also consider visiting local job fairs occasionally.

Technologies: One of the most common questions asked by beginning networkers is "Which technology should I focus on first? Microsoft? UNIX? Cisco? Novell?" As with certifications, preferences vary from company to company and person to person.

One way for a person to answer this question is to start with the technology that appears most interesting to them personally. Researching a company that one plans to interview with and choosing a technology that the company deems important, is another way. Ultimately it probably matters little which networking technology one learns first. More importantly, one should acknowledge that technology changes rapidly and that the person who can enjoy a successful career by learning about only one technology is rare indeed.

Focus on the Basics: Computer networking involves a certain number of fundamental technologies. These technologies form the basis of many networking courses. Regardless of the form of education one chooses to invest in, one's career will always benefit from deeper study of technologies like IP and TCP/IP, the OSI model, Ethernet, internet working and others listed on this site, whether through formal coursework or through self-study.

Conclusion: Some people have asserted that networking (and IT generally) is a "young person's game," and that companies generally prefer to turn over their employee base periodically, to bring in younger, more affordable workers. This concept might sound appealing to some, but if it were true, it would make networking careers less inviting to most people. Realistically, the field of computer networking presents so much complexity and involves such a wide range of technologies, that most serious companies should value both experienced employees and ambitious new employees highly. In fact, an effective career strategy involves seeking out more experienced people in one's field and learning new skills from these *mentors*.

Many firms view four-years degrees as a sign of commitment to the field. Network technology changes very fast, so employers care both about a person's current knowledge and also their ability to learn and adapt for the future. Certifications effectively prove current knowledge, but college degrees best demonstrate one's general learning ability.

Self-study in networking is always effective and underrated by many. By making contacts with those in networking careers, either people in one's local area, or individuals or sites on the Internet, one can quickly acquire a wealth of information ranging from technical details, to advice on writing a resume, to advice on specific hiring companies, schools and so on.

Management Information Systems (MIS)

With offices in 85 countries and with more than 1,500 employees, our systems need to be fast, effective and comprehensive. Alltech is all about communication. Information Technology has been a key enabler in facilitating the rapid growth of the company. One of our key functions is the development of effective information systems to satisfy our internal and external demands. Dealing with the challenges of the information age requires a dynamic team of individuals working in concert around the world.

Carlos joined Alltech as MIS Manager for Latin America in 2000. Based in Kentucky, he manages eleven different countries, driving all of them through a period of growth. Carlos ensures that each of them possess the same high-level technological infrastructure required to support the continued international growth of Alltech. During the last three years he has been afforded the opportunity to redesign the networks and update the equipment within all of the countries for which he is responsible. Two of these countries, Brazil and Mexico, represent some of Alltech's biggest sales markets and through Alltech MIS, Carlos works with and manages some excellent IT professionals. In each of the other nine countries, he manages relationships with local companies that provide Alltech with front-line IT support.

Industry Organizations

World Information Technology and Services Alliance (WITSA) is a consortium of over 60 information technology (IT) industry associations from economies around the world. Founded in 1978 and originally known as the World Computing Services Industry Association, WITSA has increasingly assumed an active advocacy role in international public policy issues affecting the creation of a robust global information infrastructure.

The Information Technology Association of America (ITAA) is an industry trade group for several U.S. information technology companies.

Founded in 1961 as the Association of Data Processing Services Organizations (ADAPSO), the Information Technology Association of America (ITAA) provides global public policy, business networking and national leadership to promote the continued rapid growth of the IT industry. ITAA consists of approximately 325 corporate members throughout the U.S. and is secretariat of the World Information Technology and Services Alliance (WITSA), a global network of 67 countries' IT associations.

NASSCOM is the Industry body/chamber of commerce for Information Technology Companies in India. It has been instrumental in helping the Indian Information Technology Industry grow.

Career Profile: Application Architect

Application Architects design, construct, deploy and maintain applications that target a specific business need. Those in this career design and implement business object models and application services, determine the requirements for these services and identify the techniques and technologies that need to be applied to help define a system and its operational requirements. They must be able to identify the types, organization and interactions of the software components that will be responsible for delivering the defined user services.

Information Technology: HSBC in Hong Kong has been pursuing an aggressive business strategy, particularly in Personal Financial Services, Commercial Banking and Wealth Management HSBC in Hong Kong has been pursuing an aggressive business strategy, particularly in Personal Financial Services, Commercial Banking and Wealth Management. As IT is a key enabler of this strategy, the demand for IT development resources to support its development and implementation has increased significantly.

HTSA (HSBC Technology Services Asia-Pacific) is made up of over 30 major departments, with about 2,800 staff. These departments increase the operational efficiency of HSBC in the Asia-Pacific region and improve and increase the range of services offered to customers through the design, development, installation, operation and maintenance of appropriate computer systems. Each department is responsible for a distinct area of systems development or support. The departments are grouped by types of equipment/applications and by end-users served.

IT Development: IT Development is responsible for the design and development of Banking Systems and Group Systems for Personal Financial Services and Commercial Banking and also looks after the deployment of Enterprise Data. To ensure that the systems meet the unique requirements of our business, it is our best practice to deploy Group Systems or develop our major software applications and systems internally rather than buying third-party packages. To ensure that our applications are efficiently and professionally constructed, we have invested heavily in the use of modern software tools. At the same time, we encourage a full understanding of the technical environment. Where efficiency is paramount, assembly languages and other low level tools are employed.

Programming languages we use include COBOL, PL/1, RPG, SQL, C, C++, JAVA, Visual Basic and Assembler and control languages JCL, CL, DCL and REXX. Systems we adopt include z/OS, OS/390, TSO, CICS, OS/400, OS/2, Windows, UNIX and CWD. We also employ DB2, IDMS, DB/400, SQL Server, Sybase and ORACLE.

IT Operations: IT Operations comprise Computer Operations, Infrastructure and Telecommunications teams. Our mainframe computers are IBM and IBM-compatibles. We use IBM iSeries mid-range and mini-computers. We use IBM RS/6000, HP, Domino and MS Windows NT servers. Our Microcomputers and Terminals are IBM PC/Compatibles/LAN, OS/2 Banking Terminals, NCR ATMs, Unix/ Windows NT Workstations and NCR Cash Deposit Machine. In our network, the major components are Nortel Passport/Cisco Routers. Our Group Router Network (GRN) covers the major/larger offices, whilst Global Data Network (GDN) covers the branch networks globally and comprises over 10,000 Cisco routers, in 76 countries and territories.

IT General: Other IT functional areas include IT Architecture, Information Security, IT Quality, Finance and Planning. These areas serve the important functions of aligning business strategies with emerging technologies, facilitating the implementation of Group IT security standards, improving transparency of IT charges and aligning quality management practices across Asia-Pacific regions and Group standards.

Corporate, Investment Banking and Markets (CIBM) IT HK

Global CIBM IT provides software solutions to our Corporate, Investment Banking and Markets business globally. This is a complex and fast-paced business, involving high-value transactions that contribute to one-third of the Bank's total profits. CIBM IT HK is responsible for supporting the CIBM business in Hong Kong and overseeing IT activities for CIBM businesses in over 30 entities across more than 20 Asia-Pacific countries and territories, as well as working closely with our other regional CIBM IT centres in London, France and New York. CIBM IT prides itself on the early adoption of advanced new technologies, fostering an environment of continuous development.

Those who excel within CIBM IT in HSBC should be willing to face challenges in a demanding, high-pressure role and possess strong problem solving abilities, a highly analytical mindset and an enthusiasm for keeping abreast of emerging technologies. Far from a stereotypical and sedentary programming function, our people often travel globally and coordinate with our product teams in Asia Pacific and other regions worldwide.

IT Graduate Opportunities

Training and Development: People are our most valuable resources. The enormous range of knowledge and expertise our staff possess is absolutely vital to the very existence and future of our organisation. IT therefore ensures that this knowledge and expertise are carefully nurtured and the personal needs of our people are recognised and met. With our emphasis on teamwork and on ensuring that we meet our user needs, we look for people who are good team players and are able to communicate effectively with both technical and non-technical people. Graduates majoring in IT- and computer-related studies who have strong analytical skills and good attention to detail must, above all, have a strong commitment to a career at the leading edge of the IT profession. For these people, our policy to fill senior positions by promotion from within means that the opportunities for career advancement are excellent. Training plays a key role. Initial training consists of a one-month induction programme, comprising full-time classroom lectures delivered by experienced staff, various video courses and course projects to practise teamwork and to familiarise staff with the development environment. Staff can continue learning with classroom training, self-study training and on-the-job training. IT PATH courses in Information Processing, Banking Applications, Technical Skills and Human Resources Development are available to help employees as their careers progress.

Programmers with a higher diploma can progress to Systems Analyst/Programmer positions after 12 months. After 24 months' initial training, Systems Analysts and Programmers who hold a degree become eligible for promotion to IT Officer. As IT Officers progress and develop, they take increasing responsibility for analysis, design and project/ team leadership on more complex systems, up to and possibly beyond department management.

Benefits and Applications: HSBC offers job security, prospective career development, a pleasant working environment and attractive fringe benefits. Among our benefits are: Housing Loan Scheme, Medical Insurance, Variable Incentives/Bonus-related Share Options Scheme, Savings-related Share Options Scheme, Local Staff Defined Contribution Retirement Scheme/MPF, 22 days' Annual Leave and Credit Facilities (including overdrafts and credit cards).

To Apply: Graduates with little or no working experience who are interested in entry-level positions, please download and complete the application form as a PDF file or Word document.

IT Development: IT Development is responsible for the design and development of Banking Systems and Group Systems for Personal Financial Services and Commercial Banking and also looks after the deployment of Enterprise Data. To ensure that the systems meet the unique requirements of our business, it is our best practice to deploy Group Systems or develop our major software applications and systems internally rather than buying third-party packages.

To ensure that our applications are efficiently and professionally constructed, we have invested heavily in the use of modern software tools. At the same time, we encourage a full understanding of the technical environment. Where efficiency is paramount, assembly languages and other low level tools are employed.

Programming languages we use include COBOL, PL/1, RPG, SQL, C, C++, JAVA, Visual Basic and Assembler and control languages JCL, CL, DCL and REXX. Systems we adopt include z/OS, OS/390, TSO, CICS, OS/400, OS/2, Windows, UNIX and CWD. We also employ DB2, IDMS, DB/400, SQL Server, Sybase and ORACLE.

IT Operations: IT Operations comprise Computer Operations, Infrastructure and Telecommunications teams. Our mainframe computers are IBM and IBM-compatibles. We use IBM iSeries mid-range and mini-computers. We use IBM RS/6000, HP, Domino and MS Windows NT servers. Our Microcomputers and Terminals are IBM PC/Compatibles/LAN, OS/2 Banking Terminals, NCR ATMs, Unix/

Windows NT Workstations and NCR Cash Deposit Machine. In our network, the major components are Nortel Passport/Cisco Routers. Our Group Router Network (GRN) covers the major/larger offices, whilst Global Data Network (GDN) covers the branch networks globally and comprises over 10,000 Cisco routers, in 76 countries and territories.

IT General: Other IT functional areas include IT Architecture, Information Security, IT Quality, Finance and Planning. These areas serve the important functions of aligning business strategies with emerging technologies, facilitating the implementation of Group IT security standards, improving transparency of IT charges and aligning quality management practices across Asia-Pacific regions and Group standards.

Training and Professional Qualifications

On-the-job training and in-house courses are typically provided to help you develop technical and business skills, though external courses may be necessary to familiarise you with specific programming languages and business software. Depending on your role, you may also be offered training in project management, or in softer skills such as communication, team leading and presentations. The British Computer Society (BCS) offers professional qualifications. e-skills UK—The Sector Skills Council for IT and Telecoms has also launched the Graduate Professional Development Award (GPDA), which combines National Vocational Qualification (NVQ) and key skill units to complement honours degrees.

Rewards: Salaries for IT professionals typically start in the region of £19,000—£25,000, progressing to £30,000—£45,000 for those at senior level with experience (for example, after 10-15 years). Salaries for senior analysts and programmers may be higher in this, the financial sector and in London, the South East and the Midlands, where they could top £80,000. Individuals with good business skills who move into more strategic business development roles are also likely to command higher salaries.

Prospects: Career progression for most in IT roles is mainly into management via team leadership and project management. You may choose to remain in one area of technical specialisation or to broaden your career options into other IT, business development, customer-related or management functions.

Computer Programmers

Significant Points: Sixty-seven percent of computer programmers held a college or higher degree in 2004; nearly half held a bachelor's degree and about 1 in 5 held a graduate degree. Employment is expected to grow much more slowly than that for other computer specialists.

Prospects likely will be best for college graduates with knowledge of a variety of programming languages and tools; those with less formal education or its equivalent in work experience are apt to face strong competition for programming jobs.

Nature of the Work: Computer programmers write, test and maintain the detailed instructions, called programmes, that computers must follow to perform their functions. Programmers also conceive, design and test logical structures for solving problems by computer. Many technical innovations in programming—advanced computing technologies and sophisticated new languages and programming tools—have redefined the role of a programmer and elevated much of the programming work done today. Job titles and descriptions may vary, depending on the organization. In this occupational statement, *computer programmers* are individuals whose main job function is programming; this group has a wide range of responsibilities and educational backgrounds.

Computer programmes tell the computer what to do—which information to identify and access, how to process it and what equipment to use. Programmes vary widely depending on the type of information to be accessed or generated. For example, the instructions involved in updating financial records are very different from those required to duplicate conditions on an aircraft for pilots training in a flight simulator. Although simple programmes can be written in a few hours, programmes that use complex mathematical formulas whose solutions can only be approximated or that draw data from many existing systems may require more than a year of work. In most cases, several programmers work together as a team under a senior programmer's supervision.

Programmers write programmes according to the specifications determined primarily by computer software engineers and systems analysts. After the design process is complete, it is the job of the programmer to convert that design into a logical series of instructions that the computer can follow. The programmer codes these instructions in a conventional programming language such as COBOL; an artificial intelligence language such as Prolog; or one of the most advanced object-oriented languages, such as Java, C++, or ACTOR. Different programming languages are used depending on the purpose of the programme. COBOL, for example, is commonly used for business applications, whereas Fortran (short for "formula translation") is used in science and engineering. C++ is widely used for both scientific and business applications. Extensible Markup Language (XML) has become a popular programming tool for Web programmers, along with J2EE (Java 2 Platform). Programmers generally know more than one programming language and, because many languages are similar, they often can learn new languages relatively easily.

In practice, programmers often are referred to by the language they know, such as Java programmers, or by the type of function they perform or environment in which they work—for example, database programmers, mainframe programmers, or Web programmers.

Many programmers update, repair, modify and expand existing programmes. When making changes to a section of code, called a routine, programmers need to make other users aware of the task that the routine is to perform. They do this by inserting comments in the coded instructions so that others can understand the programme. Many programmers use computer-assisted software engineering (CASE) tools to automate much of the coding process. These tools enable a programmer to concentrate on writing the unique parts of the programme, because the tools automate various pieces of the programme being built. CASE tools generate whole sections of code automatically, rather than line by line. Programmers also use libraries of basic code that can be modified or customized for a specific application. This approach yields more reliable and consistent programmes and increases programmers' productivity by eliminating some routine steps. Programmers test a programme by running it to ensure that the instructions are correct and that the programme produces the desired outcome. If errors do occur, the programmer must make the appropriate change and re-check the programme until it produces the correct results.

This process is called testing and debugging. Programmers may continue to fix these problems throughout the life of a programme. Programmers working in a mainframe environment, which involves a large centralized computer, may prepare instructions for a computer operator who will run the programme. Programmers also may contribute to a manual for persons who will be using the programme.

Computer programmers often are grouped into two broad types—applications programmers and systems programmers. *Applications programmers* write programmes to handle a specific job, such as a programme to track inventory within an organization. They also may revise existing packaged software or customize generic applications which are frequently purchased from vendors.

Systems programmers, in contrast, write programmes to maintain and control computer systems software, such as operating systems, networked systems and database systems. These workers make changes in the instructions that determine how the network, workstations and central processing unit of the system handle the various jobs they have been given and how they communicate with peripheral equipment such as terminals, printers and disk drives.

Because of their knowledge of the entire computer system, systems programmers often help applications programmers determine the source of problems that may occur with their programmes.

Programmers in software development companies may work directly with experts from various fields to create software—either programmes designed for specific clients or packaged software for general use—ranging from games and educational software to programmes for desktop publishing and financial planning. Programming of packaged software constitutes one of the most rapidly growing segments of the computer services industry.

In some organizations, particularly small ones, workers commonly known as *programmer-analysts* are responsible for both the systems analysis and the actual programming work. Advanced programming languages and new object-oriented programming capabilities are increasing the efficiency and productivity of both programmers and users. The transition from a mainframe environment to one that is based primarily on personal computers (PCs) has blurred the once rigid distinction between the programmer and the user. Increasingly, adept end users are taking over many of the tasks previously performed by programmers. For example, the growing use of packaged software, such as spreadsheet and database management software packages, allows users to write simple programmes to access data and perform calculations.

Working Conditions: Programmers generally work in offices in comfortable surroundings. Many programmers may work long hours or weekends to meet deadlines or fix critical problems that occur during off hours. Telecommuting is becoming common for a wide range of computer professionals, including computer programmers. As computer networks expand, more programmers are able to make corrections or fix problems remotely using modems, e-mail and the Internet to connect to a customer's computer.

Like other workers who spend long periods in front of a computer terminal typing at a keyboard, programmers are susceptible to eyestrain, back discomfort and hand and wrist problems such as carpal tunnel syndrome.

Training, Other Qualifications and Advancement: Although there are many training paths available for programmers, mainly because employers' needs are so varied, the level of education and experience employers seek has been rising due to the growing number of qualified applicants and the specialization involved with most programming tasks. Bachelor's degrees are commonly required, although some programmers may qualify for certain jobs with 2-year degrees or certificates. The associate degree is a widely used entry-level credential for prospective computer programmers. Most community colleges and many independent technical institutes and proprietary schools offer an associate degree in computer science or a related information technology field.

Employers primarily are interested in programming knowledge and computer programmers can become certified in a programming language such as C++ or Java. College graduates who are interested in changing careers or developing an area of expertise also may return to a 2-year community college or technical school for additional training. In the absence of a degree, substantial specialized experience or expertise may be needed. Even when hiring programmers with a degree, employers appear to place more emphasis on previous experience.

Some computer programmers hold a college degree in computer science, mathematics, or information systems, whereas others have taken special courses in computer programming to supplement their degree in a field such as accounting, inventory control, or another area of business. As the level of education and training required by employers continues to rise, the proportion of programmers with a college degree should increase in the future. As indicated by the following tabulation, more than two-thirds of computer programmers had a bachelor's or higher degree in 2004.

High school graduate or less	8.3%
Some college, no degree	14.1
Associate degree	10.2
Bachelor's degree	49.1
Graduate degree	18.3

Required skills vary from job to job, but the demand for various skills generally is driven by changes in technology. Employers using computers for scientific or engineering applications usually prefer college graduates who have degrees in computer or information science, mathematics, engineering, or the physical sciences. Graduate degrees in related fields are required for some jobs. Employers who use computers for business applications prefer to hire people who have had college courses in management information systems and business and who possess strong programming skills.

Although knowledge of traditional languages still is important, employers are placing increasing emphasis on newer, object-oriented programming languages and tools such as C++ and Java. Additionally, employers are seeking persons familiar with fourth-generation and fifth-generation languages that involve graphic user interface and systems programming. Employers also prefer applicants who have general business skills and experience related to the operations of the firm. Students can improve their employment prospects by participating in a college work-study programme or by undertaking an internship.

Most systems programmers hold a 4-year degree in computer science. Extensive knowledge of a variety of operating systems is essential for such workers. This includes being able to configure an operating system to work with different types of hardware and having the skills needed to adapt the operating system to best meet the needs of a particular organization. Systems programmers also must be able to work with database systems, such as DB2, Oracle, or Sybase.

When hiring programmers, employers look for people with the necessary programming skills who can think logically and pay close attention to detail. The job calls for patience, persistence and the ability to work on exacting analytical work, especially under pressure. Ingenuity and creativity are particularly important when programmers design solutions and test their work for potential failures. The ability to work with abstract concepts and to do technical analysis is especially important for systems programmers because they work with the software that controls the computer's operation. Because programmers are expected to work in teams and interact directly with users, employers want programmers who are able to communicate with nontechnical personnel.

Entry-level or junior programmers may work alone on simple assignments after some initial instruction, or they may be assigned to work on a team with more experienced programmers. Either way, beginning programmers generally must work under close supervision. Because technology changes so rapidly, programmers must continuously update their knowledge and skills by taking courses sponsored by their employer or by software vendors, or offered through local community colleges and universities.

For skilled workers who keep up to date with the latest technology, the prospects for advancement are good. In large organizations, programmers may be promoted to lead programmer and be given supervisory responsibilities. Some applications programmers may move into systems programming after they gain experience and take courses in systems software. With general business experience, programmers may become programmer-analysts or systems analysts or be promoted to managerial positions. Other programmers, with specialized knowledge and experience with a language or operating system, may work in research and development for multimedia or Internet technology and may even become computer software engineers. As employers increasingly contract with outside firms to do programming jobs, more opportunities should arise for experienced programmers with expertise in a specific area to work as consultants.

Certification is a way to demonstrate a level of competence and may provide a jobseeker with a competitive advantage. In addition to language-specific certificates that a programmer can obtain, product vendors or software firms also offer certification and may require professionals who work with their products to be certified. Voluntary certification also is available through various other organizations.

Employment: Computer programmers held about 455,000 jobs in 2004. Programmers are employed in almost every industry, but the largest concentration is in computer systems design and related services. Large numbers of programmers also work for telecommunications companies, software publishers, financial institutions, insurance carriers, educational institutions and government agencies. Many computer programmers are employed on a temporary or contract basis or work as independent consultants, providing companies expertise with new programming languages or specialized areas of application.

Rather than hiring programmers as permanent employees and then laying them off after a job is completed, employers can contract with temporary help agencies, with consulting firms, or with programmers themselves. A marketing firm, for example, may require programming services only to write and debug the software necessary to get a new customer database running. Bringing in an independent contractor or consultant with experience in a new or advanced programming language enables the firm to complete the job without having to retrain existing workers. Such jobs may last anywhere from several weeks to a year or longer. There were 25,000 self-employed computer programmers in 2004.

Job Outlook: As programming tasks become increasingly sophisticated and additional levels of skill and experience are demanded by employers, graduates of 2-year programmes and people with less than a 2-year degree or its equivalent in work experience will face strong competition for programming jobs. Competition for entry-level positions, however, also can affect applicants with a bachelor's degree.

Prospects should be best for college graduates with knowledge of and experience working with, a variety of programming languages and tools—including C++ and other object-oriented languages such as Java, as well as newer, domain-specific languages that apply to computer networking, database management and Internet application development. Obtaining vendor-specific or language-specific certification also can provide a competitive edge.

Because demand fluctuates with employers' needs, jobseekers should keep up to date with the latest skills and technologies. Individuals who want to become programmers can enhance their prospects by combining the appropriate formal training with practical work experience.

Employment of programmers is expected to grow more slowly than the average for all occupations through the year 2014. Sophisticated computer software now has the capability to write basic code, eliminating the need for many programmers to do this routine work. The consolidation and centralization of systems and applications, developments in packaged software, advances in programming languages and tools and the growing ability of users to design, write and implement more of their own programmes mean that more of the programming functions can be transferred from programmers to other types of information workers, such as computer software engineers.

Another factor limiting growth in employment is the outsourcing of these jobs to other countries. Computer programmers can perform their job function from anywhere in the world and can digitally transmit their programmes to any location via e-mail. Programmers are at a much higher risk of having their jobs outsourced abroad than are workers involved in more complex and sophisticated information technology functions, such as software engineering, because computer programming has become an international language, requiring little localized or specialized knowledge. Additionally, the work of computer programmers can be routinized, once knowledge of a particular programming language is mastered. Nevertheless, employers will continue to need programmers who have strong technical skills and who understand an employer's business and its programming requirements. This means that programmers will have to keep abreast of changing programming languages and techniques. Given the importance of networking and the expansion of client/server, Web-based and wireless environments, organizations will look for programmers who can support data communications and help implement electronic commerce and intranet strategies.

Demand for programmers with strong object-oriented programming capabilities and technical specialization in areas such as client/server programming, wireless applications, multimedia technology and graphic user interface likely will stem from the expansion of intranets, extranets and Internet applications. Programmers also will be needed to create and maintain expert systems and embed these technologies in more products. Finally, a growing emphasis on cybersecurity will lead to increased demand for programmers who are familiar with digital security issues and skilled in using appropriate security technology.

Jobs for both systems and applications programmers should be most plentiful in data-processing service firms, software houses and computer consulting businesses. These types of establishments are part of computer systems design and related services and software publishers, which are projected to be among the fastest growing industries in the economy over the 2004-14 period.

As organizations attempt to control costs and keep up with changing technology, they will need programmers to assist in conversions to new computer languages and systems. In addition, numerous job openings will result from the need to replace programmers who leave the labour force or transfer to other occupations such as manager or systems analyst.

Earnings: Median annual earnings of computer programmers were $62,890 in May 2004. The middle 50 percent earned between $47,580 and $81,280 a year. The lowest 10 percent earned less than $36,470; the highest 10 percent earned more than $99,610. Median annual earnings in the industries employing the largest numbers of computer programmers in May 2004 are shown below:

Software publishers	$73,060
Computer systems design and related services	67,600
Data processing, hosting and related services	64,540
Insurance carriers	62,990
Management of companies and enterprises	62,160

According to the National Association of Colleges and Employers, starting salary offers for graduates with a bachelor's degree in computer science averaged $50,820 a year in 2005.

According to Robert Half International, a firm providing specialized staffing services, average annual starting salaries in 2005 ranged from $52,500 to $83,250 for applications development programmers/analysts and from $55,000 to $88,250 for software developers. Average starting salaries for mainframe systems programmers ranged from $50,250 to $67,500 in 2005.

Computer Software Engineers

Significant Points: Computer software engineers are projected to be one of the fastest growing occupations over the 2004-14 period.

Very good opportunities are expected for college graduates with at least a bachelor's degree in computer engineering or computer science and with practical work experience.

Computer software engineers must continually strive to acquire new skills in conjunction with the rapid changes that are occurring in computer technology.

Nature of the Work: The explosive impact of computers and information technology on our everyday lives has generated a need to design and develop new computer software systems and to incorporate new technologies into a rapidly growing range of applications. The tasks performed by workers known as computer software engineers evolve quickly, reflecting new areas of specialization or changes in technology, as well as the preferences and practices of employers. Computer software engineers apply the principles and techniques of computer science, engineering and mathematical analysis to the design, development, testing and evaluation of the software and systems that enable computers to perform their many applications.

Software engineers working in applications or systems development analyze users' needs and design, construct, test and maintain computer applications software or systems. Software engineers can be involved in the design and development of many types of software, including software for operating systems and network distribution and compilers, which convert programmes for execution on a computer. In programming, or coding, software engineers instruct a computer, line by line, how to perform a function. They also solve technical problems that arise. Software engineers must possess strong programming skills, but are more concerned with developing algorithms and analyzing and solving programming problems than with actually writing code.

Computer applications software engineers analyze users' needs and design, construct and maintain general computer applications software or specialized utility programmes. These workers use different programming languages, depending on the purpose of the programme.

The programming languages most often used are C, C++ and Java, with Fortran and COBOL used less commonly. Some software engineers develop both packaged systems and systems software or create customized applications.

Computer systems software engineers coordinate the construction and maintenance of a company's computer systems and plan their future growth. Working with the company, they coordinate each department's computer needs—ordering, inventory, billing and payroll recordkeeping, for example—and make suggestions about its technical direction. They also might set up the company's intranets—networks that link computers within the organization and ease communication among the various departments.

Systems software engineers work for companies that configure, implement and install complete computer systems. These workers may be members of the marketing or sales staff, serving as the primary technical resource for sales workers and customers.

They also may be involved in product sales and in providing their customers with continuing technical support. Since the selling of complex computer systems often requires substantial customization for the purchaser's organization, software engineers help to explain the requirements necessary for installing and operating the new system in the purchaser's computing environment. In addition, systems software engineers are responsible for ensuring security across the systems they are configuring.

Computer software engineers often work as part of a team that designs new hardware, software and systems. A core team may comprise engineering, marketing, manufacturing and design people, who work together until the product is released.

Working Conditions: Computer software engineers normally work in well-lighted and comfortable offices or laboratories in which computer equipment is located. Most software engineers work at least 40 hours a week; however, due to the project-oriented nature of the work, they also may have to work evenings or weekends to meet deadlines or solve unexpected technical problems. Like other workers who sit for hours at a computer, typing on a keyboard, software engineers are susceptible to eyestrain, back discomfort and hand and wrist problems such as carpal tunnel syndrome.

As they strive to improve software for users, many computer software engineers interact with customers and coworkers. Computer software engineers who are employed by software vendors and consulting firms, for example, spend much of their time away from their offices, frequently travelling overnight to meet with customers. They call on customers in businesses ranging from manufacturing plants to financial institutions.

As networks expand, software engineers may be able to use modems, laptops, e-mail and the Internet to provide more technical support and other services from their main office, connecting to a customer's computer remotely to identify and correct developing problems.

Training, Other Qualifications and Advancement: Most employers prefer to hire persons who have at least a bachelor's degree and broad knowledge of and experience with, a variety of computer systems and technologies. The usual degree concentration for applications software engineers is computer science or software engineering; for systems software engineers, it is computer science or computer information systems. Graduate degrees are preferred for some of the more complex jobs.

Academic programmes in software engineering emphasize software and may be offered as a degree option or in conjunction with computer science degrees. Increasing emphasis on computer security suggests that software engineers with advanced degrees that include mathematics and systems design will be sought after by software developers, government agencies and consulting firms specializing in information assurance and security. Students seeking software engineering jobs enhance their employment opportunities by participating in internship or co-op programmes offered through their schools.

These experiences provide the students with broad knowledge and experience, making them more attractive candidates to employers. Inexperienced college graduates may be hired by large computer and consulting firms that train new employees in intensive, company-based programmes. In many firms, new hires are mentored and their mentors have an input into the performance evaluations of these new employees.

For systems software engineering jobs that require workers who have a college degree, a bachelor's degree in computer science or computer information systems is typical. For systems engineering jobs that place less emphasis on workers having a computer-related degree, computer training programmes leading to certification are offered by systems software vendors. Nonetheless, most training authorities feel that programme certification alone is not sufficient for the majority of software engineering jobs.

Persons interested in jobs as computer software engineers must have strong problem-solving and analytical skills. They also must be able to communicate effectively with team members, other staff and the customers they meet. Because they often deal with a number of tasks simultaneously, they must be able to concentrate and pay close attention to detail.

As is the case with most occupations, advancement opportunities for computer software engineers increase with experience. Entry-level computer software engineers are likely to test and verify ongoing designs. As they become more experienced, they may become involved in designing and developing software. Eventually, they may advance to become a project manager, manager of information systems, or chief information officer. Some computer software engineers with several years of experience or expertise find lucrative opportunities working as systems designers or independent consultants or starting their own computer consulting firms.

As technological advances in the computer field continue, employers demand new skills. Computer software engineers must continually strive to acquire such skills if they wish to remain in this extremely dynamic field. For example, computer software engineers interested in working for a bank should have some expertise in finance as they integrate new technologies into the computer system of the bank. To help them keep up with the changing technology, continuing education and professional development seminars are offered by employers, software vendors, colleges and universities, private training institutions and professional computing societies.

Employment: Computer software engineers held about 800,000 jobs in 2004. Approximately 460,000 were computer applications software engineers and around 340,000 were computer systems software engineers. Although they are employed in most industries, the largest concentration of computer software engineers—almost 30 percent—are in computer systems design and related services. Many computer software engineers also work for establishments in other industries, such as software publishers, government agencies, manufacturers of computers and related electronic equipment and management of companies and enterprises.

Employers of computer software engineers range from start-up companies to established industry leaders. The proliferation of Internet, e-mail and other communications systems is expanding electronics to engineering firms that are traditionally associated with unrelated disciplines. Engineering firms specializing in building bridges and powerplants, for example, hire computer software engineers to design and develop new geographic data systems and automated drafting systems. Communications firms need computer software engineers to tap into growth in the personal communications market. Major communications companies have many job openings for both computer software applications engineers and computer systems engineers. An increasing number of computer software engineers are employed on a temporary or contract basis, with many being self-employed, working independently as consultants. Some consultants work for firms that specialize in developing and maintaining client companies' Web sites and intranets. About 23,000 computer software engineers were self-employed in 2004.

Job Outlook: Computer software engineers are projected to be one of the fastest-growing occupations from 2004 to 2014. Rapid employment growth in the computer systems design and related services industry, which employs the greatest number of computer software engineers, should result in very good opportunities for those college graduates with at least a bachelor's degree in computer engineering or computer science and practical experience working with computers. Employers will continue to seek computer professionals with strong programming, systems analysis, interpersonal and business skills. With the software industry beginning to mature, however and with routine software engineering work being increasingly outsourced overseas, job growth will not be as rapid as during the previous decade.

Employment of computer software engineers is expected to increase much faster than the average for all occupations, as businesses and other organizations adopt and integrate new technologies and seek to maximize the efficiency of their computer systems. Competition among businesses will continue to create an incentive for increasingly sophisticated technological innovations and organizations will need more computer software engineers to implement these changes. In addition to jobs created through employment growth, many job openings will result annually from the need to replace workers who move into managerial positions, transfer to other occupations, or leave the labour force.

Demand for computer software engineers will increase as computer networking continues to grow. For example, the expanding integration of Internet technologies and the explosive growth in electronic commerce—doing business on the Internet—have resulted in rising demand for computer software engineers who can develop Internet, intranet and World Wide Web applications. Likewise, expanding electronic data-processing systems in business, telecommunications, government and other settings continue to become more sophisticated and complex.

Growing numbers of systems software engineers will be needed to implement, safeguard and update systems and resolve problems. Consulting opportunities for computer software engineers also should continue to grow as businesses seek help to manage, upgrade and customize their increasingly complicated computer systems.

New growth areas will continue to arise from rapidly evolving technologies. The increasing uses of the Internet, the proliferation of Web sites and mobile technology such as the wireless Internet have created a demand for a wide variety of new products. As individuals and businesses rely more on hand-held computers and wireless networks, it will be necessary to integrate current computer systems with this new, more mobile technology.

Also, information security concerns have given rise to new software needs. Concerns over "cybersecurity" should result in businesses and government continuing to invest heavily in software that protects their networks and vital electronic infrastructure from attack. The expansion of this technology in the next 10 years will lead to an increased need for computer engineers to design and develop the software and systems to run these new applications and integrate them into older systems.

As with other information technology jobs, employment growth of computer software engineers may be tempered somewhat as more software development is contracted out abroad. Firms may look to cut costs by shifting operations to lower wage foreign countries with highly educated workers who have strong technical skills. At the same time, jobs in software engineering are less prone to being sent abroad compared with jobs in other computer specialties, because the occupation requires innovation and intense research and development.

Earnings: Median annual earnings of computer applications software engineers who worked full time in May 2004 were about $74,980. The middle 50 percent earned between $59,130 and $92,130. The lowest 10 percent earned less than $46,520 and the highest 10 percent earned more than $113,830. Median annual earnings in the industries employing the largest numbers of computer applications software engineers in May 2004 were as follows:

Software publishers	$79,930
Management, scientific and technical consulting services	78,460
Computer systems design and related services	76,910
Management of companies and enterprises	70,520
Insurance carriers	68,440

Median annual earnings of computer systems software engineers who worked full time in May 2004 were about $79,740. The middle 50 percent earned between $63,150 and $98,220. The lowest 10 percent earned less than $50,420 and the highest 10 percent earned more than $118,350. Median annual earnings in the industries employing the largest numbers of computer systems software engineers in May 2004 are as follows:

Scientific research and development services	$91,390
Computer and peripheral equipment manufacturing	87,800
Software publishers	83,670
Computer systems design and related services	79,950
Wired telecommunications carriers	74,370

According to the National Association of Colleges and Employers, starting salary offers for graduates with a bachelor's degree in computer engineering averaged $52,464 in 2005; offers for those with a master's degree averaged $60,354. Starting salary offers for graduates with a bachelor's degree in computer science averaged $50,820.

According to Robert Half International, starting salaries for software engineers in software development ranged from $63,250 to $92,750 in 2005. For network engineers, starting salaries in 2005 ranged from $61,250 to $88,250.

Computer Systems Analysts

Significant Points: Employers generally prefer applicants who have at least a bachelor's degree in computer science, information science, or management information systems (MIS). Employment is expected to increase much faster than the average as organizations continue to adopt increasingly sophisticated technologies.

Job prospects are favourable.

Nature of the Work: All organizations rely on computer and information technology to conduct business and operate more efficiently. The rapid spread of technology across all industries has generated a need for highly trained workers to help organizations incorporate new technologies. The tasks performed by workers known as computer systems analysts evolve rapidly, reflecting new areas of specialization or changes in technology, as well as the preferences and practices of employers.

Computer systems analysts solve computer problems and apply computer technology to meet the individual needs of an organization. They help an organization to realize the maximum benefit from its investment in equipment, personnel and business processes. Systems analysts may plan and develop new computer systems or devise ways to apply existing systems' resources to additional operations. They may design new systems, including both hardware and software, or add a new software application to harness more of the computer's power. Most systems analysts work with specific types of systems—for example, business, accounting, or financial systems, or scientific and engineering systems—that vary with the kind of organization. Some systems analysts also are known as *systems developers* or *systems architects.*

Systems analysts begin an assignment by discussing the systems problem with managers and users to determine its exact nature. Defining the goals of the system and dividing the solutions into individual steps and separate procedures, systems analysts use techniques such as structured analysis, data modelling, information engineering, mathematical model building, sampling and cost accounting to plan the system. They specify the inputs to be accessed by the system, design the processing steps and format the output to meet users' needs. They also may prepare cost-benefit and return-on-investment analyses to help management decide whether implementing the proposed technology will be financially feasible.

When a system is accepted, systems analysts determine what computer hardware and software will be needed to set the system up. They coordinate tests and observe the initial use of the system to ensure that it performs as planned. They prepare specifications, flow charts and process diagrams for computer programmers to follow; then, they work with programmers to "debug," or eliminate, errors from the system. Systems analysts who do more in-depth testing of products may be referred to as *software quality assurance analysts.* In addition to running tests, these individuals diagnose problems, recommend solutions and determine whether programme requirements have been met.

In some organizations, *programmer-analysts* design and update the software that runs a computer. Because they are responsible for both programming and systems analysis, these workers must be proficient in both areas.

As this dual proficiency becomes more commonplace, these analysts are increasingly working with databases, object-oriented programming languages, as well as client–server applications development and multimedia and Internet technology.

One obstacle associated with expanding computer use is the need for different computer systems to communicate with each other. Because of the importance of maintaining up-to-date information—accounting records, sales figures, or budget projections, for example—systems analysts work on making the computer systems within an organization, or among organizations, compatible so that information can be shared among them. Many systems analysts are involved with "networking," connecting all the computers internally—in an individual office, department, or establishment—or externally, because many organizations rely on e-mail or the Internet. A primary goal of networking is to allow users to retrieve data from a mainframe computer or a server and use it on their desktop computer.

Systems analysts must design the hardware and software to allow the free exchange of data, custom applications and the computer power to process it all. For example, analysts are called upon to ensure the compatibility of computing systems between and among businesses to facilitate electronic commerce.

Working Conditions: Computer systems analysts work in offices or laboratories in comfortable surroundings. They usually work about 40 hours a week—the same as many other professional or office workers do. However, evening or weekend work may be necessary to meet deadlines or solve specific problems. Given the technology available today, telecommuting is common for computer professionals. As networks expand, more work can be done from remote locations through modems, laptops, electronic mail and the Internet.

Like other workers who spend long periods in front of a computer terminal typing on a keyboard, computer systems analysts are susceptible to eyestrain, back discomfort and hand and wrist problems such as carpal tunnel syndrome or cumulative trauma disorder.

Training, Other Qualifications and Advancement: Rapidly changing technology requires an increasing level of skill and education on the part of employees. Companies increasingly look for professionals with a broad background and range of skills, including not only technical knowledge, but also communication and other interpersonal skills. This shift from requiring workers to possess solely sound technical knowledge emphasizes workers who can handle various responsibilities.

While there is no universally accepted way to prepare for a job as a systems analyst, most employers place a premium on some formal college education. Relevant work experience also is very important. For more technically complex jobs, persons with graduate degrees are preferred. Many employers seek applicants who have at least a bachelor's degree in computer science, information science, or management information systems (MIS). MIS programmes usually are part of the business school or college and differ considerably from computer science programmes, emphasizing business and management-oriented course work and business computing courses. Employers are increasingly seeking individuals with a master's degree in business administration (MBA), with a concentration in information systems, as more firms move their business to the Internet.

Despite employers' preference for those with technical degrees, persons with degrees in a variety of majors find employment as system analysts. The level of education and type of training that employers require depend on their needs. One factor affecting these needs is changes in technology. Employers often scramble to find workers capable of implementing "hot" new technologies such as the wireless Internet. Those workers with formal education or experience in information security, for example, are in demand because of the growing need for their skills and services. Another factor driving employers' needs is the timeframe during which a project must be completed.

Employers usually look for people who have broad knowledge and experience related to computer systems and technologies, strong problem-solving and analytical skills and good interpersonal skills. Courses in computer science or systems design offer good preparation for a job in these computer occupations. For jobs in a business environment, employers usually want systems analysts to have business management or closely related skills, while a background in the physical sciences, applied mathematics, or engineering is preferred for work in scientifically oriented organizations.

Job seekers can enhance their employment opportunities by participating in internship or co-op programmes offered through their schools. Because many people develop advanced computer skills in a non-computer-related occupation and then transfer those skills to a computer occupation, a background in the industry in which the person's job is located, such as financial services, banking, or accounting, can be important. Others have taken computer science courses to supplement their study in fields such as accounting, inventory control, or other business areas.

Computer systems analysts must be able to think logically and have good communication skills. Because they often deal with a number of tasks simultaneously, the ability to concentrate and pay close attention to detail is important. Although these workers sometimes work independently, they frequently work in teams on large projects. They must be able to communicate effectively with computer personnel, such as programmers and managers, as well as with users or other staff who may have no technical computer background.

Systems analysts may be promoted to senior or lead systems analyst. Those who show leadership ability also can become project managers or advance into management positions such as manager of information systems or chief information officer. Workers with work experience and considerable expertise in a particular subject or a certain application may find lucrative opportunities as independent consultants or may choose to start their own computer consulting firms.

Technological advances come so rapidly in the computer field that continuous study is necessary to keep one's skills up to date. Employers, hardware and software vendors, colleges and universities and private training institutions offer continuing education. Additional training may come from professional development seminars offered by professional computing societies.

Employment: Computer systems analysts held about 487,000 jobs in 2004; about 28,000 were self-employed.

Although they are increasingly employed in every sector of the economy, the greatest concentration of these workers is in the computer systems design and related services industry. Firms in this industry provide services related to the commercial use of computers on a contract basis, including custom computer programming services; computer systems integration design services; computer facilities management services, including computer systems or data processing facilities support services for clients; and other computer services, such as disaster recovery services and software installation. Computer systems analysts are also employed by governments, insurance companies, financial institutions, Internet service providers, data processing services firms and universities.

A growing number of systems analysts are employed on a temporary or contract basis; many of these individuals are self-employed, working independently as contractors or consultants. For example, a company installing a new computer system may need the services of several systems analysts just to get the system running. Because not all of the analysts would be needed once the system is functioning, the company might contract for such employees with a temporary help agency or a consulting firm or with the systems analysts themselves.

Such jobs may last from several months up to 2 years or more. This growing practice enables companies to bring in people with the exact skills the firm needs to complete a particular project, rather than having to spend time or money training or retraining existing workers. Often, experienced consultants then train a company's in-house staff as a project develops.

Job Outlook: Employment of computer systems analysts is expected to grow much faster than the average for all occupations through the year 2014 as organizations continue to adopt and integrate increasingly sophisticated technologies. Job increases will be driven by very rapid growth in computer system design and related services, which is projected to be among the fastest growing industries in the U.S. economy. In addition, many job openings will arise annually from the need to replace workers who move into managerial positions or other occupations or who leave the labour force. Job growth will not be as rapid as during the previous decade, however, as the information technology sector begins to mature and as routine work is increasingly outsourced to lower-wage foreign countries.

Workers in the occupation should enjoy favourable job prospects. The demand for networking to facilitate the sharing of information, the expansion of client–server environments and the need for computer specialists to use their knowledge and skills in a problem-solving capacity will be major factors in the rising demand for computer systems analysts. Moreover, falling prices of computer hardware and software should continue to induce more businesses to expand their computerized operations and integrate new technologies into them. In order to maintain a competitive edge and operate more efficiently, firms will keep demanding system analysts who are knowledgeable about the latest technologies and are able to apply them to meet the needs of businesses.

Increasingly, more sophisticated and complex technology is being implemented across all organizations, which should fuel the demand for these computer occupations. There is a growing demand for system analysts to help firms maximize their efficiency with available technology. Expansion of electronic commerce—doing business on the Internet—and the continuing need to build and maintain databases that store critical information on customers, inventory and projects are fuelling demand for database administrators familiar with the latest technology. Also, the increasing importance being placed on "cybersecurity"—the protection of electronic information—will result in a need for workers skilled in information security.

The development of new technologies usually leads to demand for various kinds of workers. The expanding integration of Internet technologies into businesses, for example, has resulted in a growing need for specialists who can develop and support Internet and intranet applications. The growth of electronic commerce means that more establishments use the Internet to conduct their business online. The introduction of the wireless Internet, known as WiFi, creates new systems to be analyzed. The spread of such new technologies translates into a need for information technology professionals who can help organizations use technology to communicate with employees, clients and consumers. Explosive growth in these areas also is expected to fuel demand for analysts who are knowledgeable about network, data and communications security.

As technology becomes more sophisticated and complex, employers demand a higher level of skill and expertise from their employees. Individuals with an advanced degree in computer science or computer engineering, or with an MBA with a concentration in information systems, should enjoy favourable employment prospects. College graduates with a bachelor's degree in computer science, computer engineering, information science, or MIS also should enjoy favourable prospects for employment, particularly if they have supplemented their formal education with practical experience.

Because employers continue to seek computer specialists who can combine strong technical skills with good interpersonal and business skills, graduates with non-computer-science degrees, but who have had courses in computer programming, systems analysis and other information technology subjects, also should continue to find jobs in computer fields. In fact, individuals with the right experience and training can work in computer occupations regardless of their college major or level of formal education.

Earnings: Median annual earnings of computer systems analysts were $66,460 in May 2004. The middle 50 percent earned between $52,400 and $82,980 a year. The lowest 10 percent earned less than $41,730 and the highest 10 percent earned more than $99,180. Median annual earnings in the industries employing the largest numbers of computer systems analysts in May 2004 were:

Federal Government	$71,770
Computer systems design and related services	69,560
Management of companies and enterprises	67,230
Insurance carriers	66,840
State government	57,040

According to the National Association of Colleges and Employers, starting offers for graduates with a master's degree in computer science averaged $62,727 in 2005. Starting offers averaged $50,820 for graduates with a bachelor's degree in computer science; $46,189 for those with a degree in computer systems analysis; $44,417 for those with a degree in management information systems; and $44,775 for those with a degree in information sciences and systems. According to Robert Half International, starting salaries for systems analysts ranged from $61,500 to $82,500 in 2005.

Computer Scientists and Database Administrators

Significant Points: Education requirements range from an associate degree to a doctoral degree. Employment is expected to increase much faster than the average as organizations continue to adopt increasingly sophisticated technologies.

Job prospects are favourable.

Nature of the Work: The rapid spread of computers and information technology has generated a need for highly trained workers proficient in various job functions. These workers—computer scientists, database administrators and network systems and data communication analysts—include a wide range of computer specialists. Job tasks and occupational titles used to describe these workers evolve rapidly, reflecting new areas of specialization or changes in technology, as well as the preferences and practices of employers.

Computer scientists work as theorists, researchers, or inventors. Their jobs are distinguished by the higher level of theoretical expertise and innovation they apply to complex problems and the creation or application of new technology. Those employed by academic institutions work in areas ranging from complexity theory to hardware to programming-language design. Some work on multidisciplinary projects, such as developing and advancing uses of virtual reality, extending human-computer interaction, or designing robots. Their counterparts in private industry work in areas such as applying theory; developing specialized languages or information technologies; or designing programming tools, knowledge-based systems, or even computer games.

With the Internet and electronic business generating large volumes of data, there is a growing need to be able to store, manage and extract data effectively. *Database administrators* work with database management systems software and determine ways to organize and store data. They identify user requirements, set up computer databases and test and coordinate modifications to the computer database systems. An organization's database administrator ensures the performance of the system, understands the platform on which the database runs and adds new users to the system. Because they also may design and implement system security, database administrators often plan and coordinate security measures. With the volume of sensitive data generated every second growing rapidly, data integrity, backup systems and database security have become increasingly important aspects of the job of database administrators.

Because networks are configured in many ways, *network systems and data communications analysts* are needed to design, test and evaluate systems such as local area networks (LANs), wide area networks (WANs), the Internet, intranets and other data communications systems. Systems can range from a connection between two offices in the same building to globally distributed networks, voice mail and e-mail systems of a multinational organization. Network systems and data communications analysts perform network modelling, analysis and planning; they also may research related products and make necessary hardware and software recommendations. *Telecommunications specialists* focus on the interaction between computer and communications equipment. These workers design voice and data communication systems, supervise the installation of the systems and provide maintenance and other services to clients after the systems are installed.

The growth of the Internet and the expansion of the World Wide Web (the graphical portion of the Internet) have generated a variety of occupations related to the design, development and maintenance of Web sites and their servers. For example, *webmasters* are responsible for all technical aspects of a Web site, including performance issues such as speed of access and for approving the content of the site. *Internet developers* or *Web developers*, also called *Web designers*, are responsible for day-to-day site creation and design.

Working Conditions: Computer scientists and database administrators normally work in offices or laboratories in comfortable surroundings. They usually work about 40 hours a week—the same as many other professional or office workers do. However, evening or weekend work may be necessary to meet deadlines or solve specific problems. With the technology available today, telecommuting is common for computer professionals. As networks expand, more work can be done from remote locations through modems, laptops, electronic mail and the Internet.

Like other workers who spend long periods in front of a computer terminal typing on a keyboard, computer scientists and database administrators are susceptible to eyestrain, back discomfort and hand and wrist problems such as carpal tunnel syndrome or cumulative trauma disorder.

Training, Other Qualifications and Advancement: Rapidly changing technology requires an increasing level of skill and education on the part of employees. Companies look for professionals with an ever-broader background and range of skills, including not only technical knowledge, but also communication and other interpersonal skills. While there is no universally accepted way to prepare for a job as a network systems analyst, computer scientist, or database administrator, most employers place a premium on some formal college education. A bachelor's degree is a prerequisite for many jobs; however, some jobs may require only a 2-year degree. Relevant work experience also is very important. For more technically complex jobs, persons with graduate degrees are preferred.

For database administrator positions, many employers seek applicants who have a bachelor's degree in computer science, information science, or management information systems (MIS). MIS programmes usually are part of the business school or college and differ considerably from computer science programmes, emphasizing business and management-oriented coursework and business computing courses. Employers increasingly seek individuals with a master's degree in business administration (MBA), with a concentration in information systems, as more firms move their business to the Internet. For some network systems and data communication analysts, such as webmasters, an associate degree or certificate is sufficient, although more advanced positions might require a computer-related bachelor's degree. For computer and information scientists, a doctoral degree generally is required because of the highly technical nature of their work.

Despite employers' preference for those with technical degrees, persons with degrees in a variety of majors find employment in these occupations. The level of education and the type of training that employers require depend on their needs. One factor affecting these needs is changes in technology. Employers often scramble to find workers capable of implementing new technologies. Workers with formal education or experience in information security, for example, are in demand because of the growing need for their skills and services. Employers also look for workers skilled in wireless technologies as wireless networks and applications have spread into many firms and organizations.

Most community colleges and many independent technical institutes and proprietary schools offer an associate's degree in computer science or a related information technology field. Many of these programmes may be geared more toward meeting the needs of local businesses and are more occupation specific than are 4-year degree programmes. Some jobs may be better suited to the level of training that such programmes offer. Employers usually look for people who have broad knowledge and experience related to computer systems and technologies, strong problem-solving and analytical skills and good interpersonal skills. Courses in computer science or systems design offer good preparation for a job in these computer occupations. For jobs in a business environment, employers usually want systems analysts to have business management or closely related skills, while a background in the physical sciences, applied mathematics, or engineering is preferred for work in scientifically oriented organizations. Art or graphic design skills may be desirable for webmasters or Web developers.

Jobseekers can enhance their employment opportunities by participating in internship or co-op programmes offered through their schools. Because many people develop advanced computer skills in a non-computer occupation and then transfer those skills to a computer occupation, a background in the industry in which the person's job is located, such as financial services, banking, or accounting, can be important. Others have taken computer science courses to supplement their study in fields such as accounting, inventory control, or other business areas.

Computer scientists and database administrators must be able to think logically and have good communication skills. Because they often deal with a number of tasks simultaneously, the ability to concentrate and pay close attention to detail is important. Although these computer specialists sometimes work independently, they frequently work in teams on large projects. They must be able to communicate effectively with computer personnel, such as programmers and managers, as well as with users or other staff who may have no technical computer background.

Computer scientists employed in private industry may advance into managerial or project leadership positions. Those employed in academic institutions can become heads of research departments or published authorities in their field. Database administrators may advance into managerial positions, such as chief technology officer, on the basis of their experience managing data and enforcing security. Computer specialists with work experience and considerable expertise in a particular subject or a certain application may find lucrative opportunities as independent consultants or may choose to start their own computer consulting firms.

Technological advances come so rapidly in the computer field that continuous study is necessary to keep one's skills up to date. Employers, hardware and software vendors, colleges and universities and private training institutions offer continuing education. Additional training may come from professional development seminars offered by professional computing societies.

Certification is a way to demonstrate a level of competence in a particular field. Some product vendors or software firms offer certification and require professionals who work with their products to be certified. Many employers regard these certifications as the industry standard. For example, one method of acquiring enough knowledge to get a job as a database administrator is to become certified in a specific type of database management. Voluntary certification also is available through various organizations associated with computer specialists. Professional certification may afford a jobseeker a competitive advantage.

Employment: Computer scientists and database administrators held about 507,000 jobs in 2004, including about 66,000 who were self-employed. Employment was distributed among the detailed occupations as follows:

Network systems and data communication analysts	231,000
Database administrators	104,000
Computer and information scientists, research	22,000
Computer specialists, all other	149,000

Although they are increasingly employed in every sector of the economy, the greatest concentration of these workers is in the computer systems design and related services industry. Firms in this industry provide services related to the commercial use of computers on a contract basis, including custom computer programming services; computer systems integration design services; computer facilities management services, including computer systems or data processing facilities support services for clients; and other computer-related services, such as disaster recovery services and software installation.

Many computer scientists and database administrators are employed by Internet service providers; Web search portals; and data processing, hosting and related services firms. Others work for government, manufacturers of computer and electronic products, insurance companies, financial institutions and universities. A growing number of computer specialists, such as network and data communications analysts, are employed on a temporary or contract basis; many of these individuals are self-employed, working independently as contractors or consultants. For example, a company installing a new computer system may need the services of several network systems and data communication analysts just to get the system running. Because not all of the analysts would be needed once the system is functioning, the company might contract for such employees with a temporary help agency or a consulting firm or with the network systems analysts themselves. Such jobs may last from several months to 2 years or more. This growing practice enables companies to bring in people with the exact skills they need to complete a particular project, rather than having to spend time or money training or retraining existing workers. Often, experienced consultants then train a company's in-house staff as a project develops.

Job Outlook: Computer scientists and database administrators should continue to enjoy favourable job prospects. As technology becomes more sophisticated and complex, however, employers demand a higher level of skill and expertise from their employees. Individuals with an advanced degree in computer science or computer engineering or with an MBA with a concentration in information systems should enjoy favourable employment prospects. College graduates with a bachelor's degree in computer science, computer engineering, information science, or MIS also should enjoy favourable prospects, particularly if they have supplemented their formal education with practical experience. Because employers continue to seek computer specialists who can combine strong technical skills with good interpersonal and business skills, graduates with degrees in fields other than computer science who have had courses in computer programming, systems analysis and other information technology areas also should continue to find jobs in these computer fields. In fact, individuals with the right experience and training can work in these computer occupations regardless of their college major or level of formal education.

Computer scientists and database administrators are expected to be among the fastest growing occupations through 2014. Employment of these computer specialists is expected to grow much faster than the average for all occupations as organizations continue to adopt and integrate increasingly sophisticated technologies. Job increases will be driven by very rapid growth in computer systems design and related services, which is projected to be one of the fastest growing industries in the U.S. economy.

Job growth will not be as rapid as during the previous decade, however, as the information technology sector begins to mature and as routine work is increasingly outsourced overseas. In addition to growth, many job openings will arise annually from the need to replace workers who move into managerial positions or other occupations or who leave the labour force. The demand for networking to facilitate the sharing of information, the expansion of client–server environments and the need for computer specialists to use their knowledge and skills in a problem-solving capacity will be major factors in the rising demand for computer scientists and database administrators. Moreover, falling prices of computer hardware and software should continue to induce more businesses to expand their computerized operations and integrate new technologies into them. To maintain a competitive edge and operate more efficiently, firms will keep demanding computer specialists who are knowledgeable about the latest technologies and are able to apply them to meet the needs of businesses.

Increasingly, more sophisticated and complex technology is being implemented across all organizations, fuelling demand for computer scientists and database administrators. There is growing demand for network systems and data communication analysts to help firms maximize their efficiency with available technology.

Expansion of electronic commerce—doing business on the Internet—and the continuing need to build and maintain databases that store critical information on customers, inventory and projects are fuelling demand for database administrators familiar with the latest technology. Also, the increasing importance placed on cybersecurity—the protection of electronic information—will result in a need for workers skilled in information security.

The development of new technologies usually leads to demand for various kinds of workers. The expanding integration of Internet technologies into businesses, for example, has resulted in a growing need for specialists who can develop and support Internet and intranet applications. The growth of electronic commerce means that more establishments use the Internet to conduct their business online. The introduction of the wireless Internet, known as WiFi, creates new systems to be analyzed and new data to be administered. The spread of such new technologies translates into a need for information technology professionals who can help organizations use technology to communicate with employees, clients and consumers. Explosive growth in these areas also is expected to fuel demand for specialists who are knowledgeable about network, data and communications security.

Earnings: Median annual earnings of computer and information scientists, research, were $85,190 in May 2004. The middle 50 percent earned between $64,860 and $108,440. The lowest 10 percent earned less than $48,930 and the highest 10 percent earned more than $132,700. Median annual earnings of computer and information scientists employed in computer systems design and related services in May 2004 were $85,530. Median annual earnings of database administrators were $60,650 in May 2004. The middle 50 percent earned between $44,490 and $81,140. The lowest 10 percent earned less than $33,380 and the highest 10 percent earned more than $97,450. In May 2004, median annual earnings of database administrators employed in computer systems design and related services were $70,530 and for those in management of companies and enterprises, earnings were $65,990.

Median annual earnings of network systems and data communication analysts were $60,600 in May 2004. The middle 50 percent earned between $46,480 and $78,060. The lowest 10 percent earned less than $36,260 and the highest 10 percent earned more than $95,040. Median annual earnings in the industries employing the largest numbers of network systems and data communications analysts in May 2004 are shown below:

Wired telecommunications carriers	$65,130
Insurance carriers	64,660
Management of companies and enterprises	64,170
Computer systems design and related services	63,910
Local government	52,300

Median annual earnings of all other computer specialists were $59,480 in May 2004. Median annual earnings of all other computer specialists employed in computer systems design and related services were $57,430 and, for those in management of companies and enterprises, earnings were $68,590 in May 2004.

According to the National Association of Colleges and Employers, starting offers for graduates with a doctoral degree in computer science averaged $93,050 in 2005. Starting offers averaged $50,820 for graduates with a bachelor's degree in computer science; $46,189 for those with a degree in computer systems analysis; $44,417 for those with a degree in management information systems; and $44,775 for those with a degree in information sciences and systems.

According to Robert Half International, a firm providing specialized staffing services, starting salaries in 2005 ranged from $67,750 to $95,500 for database administrators. Salaries for networking and Internet-related occupations ranged from $47,000 to $68,500 for LAN administrators and from $51,750 to $74,520 for web developers. Starting salaries for information security professionals ranged from $63,750 to $93,000 in 2005.

Chapter 2

Internet Technology

Internet

The arrival of the Internet made communication between machines much easier and a number of open protocols and applications were developed to make use of this. Of these, E-mail was the forerunner and there can be few academics and students that do not have access to this now. E-mail has its limitations and it was the World Wide Web that really brought the world of networked computers to the general public.

The open standards of many of the technologies and the ease with which anyone could publish information encouraged participation by all and we need to remember what is about these technologies that makes them attractive when we try to deploy them for education. However, initially, a relatively small number of University lecturers adopted it for a range of teaching purposes but even fewer did more than post information about their courses or actual lecture notes - usually not modified in any way to take advantage of the strengths of the media such as hypertext.

One of the strengths and principle attractions of the Web is that it can provide authoring access to anybody and this is quite different from the one-way nature of education through CAL or any other media that predated it. The fact that the technology facilitates this does not of course mean that it will take place but then this is true of any educational forum. While far from needing programming skills, it still takes a certain amount of technophilia to publish a Web page. Creating them is trivial but actually publishing them can be tedious if the institution has not provided a simple means to do so. This is all about information rather than teaching and learning and it soon becomes obvious to any treading this path that you cannot take the people out of the learning equation entirely. Learning is about interaction and interaction with information alone is not enough.

History

The original concept for the Internet was developed in 1962 by Paul Baran of the RAND Corporation. RAND in the 1960s was a Cold War military think tank. RAND was commissioned by the US military to find a way to maintain military communication in the event of a nuclear attack. The idea was to create a network that has no central control that could be targeted in an attack. Baran laid out a plan using packet switching technology to maintain communications through a network of multiple communications paths. The idea for multiple paths allowed communications to continue in the even that any one or more sections of the network were neutralized due to a nuclear blast.

In 1968, a contract was awarded by ARPA (Advanced Research Projects Agency) to a company called BNN to build the first network model, which was then called ARPANET. The original ARPANET used a Honeywell computer to handle the network switching. The original network linked four nodes: the University of California in Santa Barbara, the University of California in Los Angeles, the University of Utah and SRI in Stanford.

A major event in the history of the Internet occurred in 1972 when Ray Tomlinson of BNN developed the first e-mail programme. Also in 1972, ARPA changed its name to DARPA (Defence Advanced Research Projects Agency). DARPA and Stanford were collaborating on the development of TCP/IP, which became the main communications protocol for the Internet. TCP/IP was a huge breakthrough because it allowed different types of computers on different networks to communicate with one another. Prior to this breakthrough, computers could only communicate with other computers on the same type of network. The name "Internet" was first used in the development of TCP. In 1976, Dr. Robert Metcalfe of Xerox's Palo Alto Research Facility developed Ethernet, which allowed data to be moved extremely fast. This breakthrough allowed local area networks (LANs) to become practical. Actually, in 1976, extremely fast meant a 56Kbps connection, which is the same speed that most narrowband dial-up connections still operate on due to the limitation of telephone wiring.

In 1983, every computer connected to ARPANET was required to use TCP/IP, which became the core Internet communications protocol. Also in 1983, the University of Wisconsin developed the Domain Name

System (DNS), which allowed TCP/IP packs to be delivered to a server representing a domain name. Because the Internet actually operates on IP addresses, domain names must be translated into IP addresses by Domain Name Servers. This made it much easier to remember addresses for different nodes on the Internet.

1984 saw the Internet split into two entities. MIL NET became focused on the needs of the military and military development projects. ARPANET became focused on advanced research. The US Department of Defence continued to fund and support both networks. Also in 1994, the first high-speed Internet backbones were installed by MCI. The new network was called NSFNET (National Science Foundation Network)

In 1985, the first.com domain name was registered by Symbolics, who still uses it today. In 1985, the Internet was strictly a text-based system. There were no images or web sites as we see today. Internet sites consisted of pages of text messages.

In 1990, the US Department of Defence replaced the old ARPANET infrastructure with the the new, high-speed NSFNET backbone. Also in 1990, Tim Berne's-Lee of CERN began working with a hypertext system that would eventually lead to the creation of the World Wide Web, for which he is best known. Tim Berners-Lee is frequently called the father of the world wide web due to his work in developing the concepts In 1992, the World Wide Web was introduced by CERN. The World Wide Web incorporated hypertext linking capabilities, a standardized HTML scripting language, standardized resource identifiers (URIs/URLs) and utilized a client-server model for requesting web pages and objects using a hypertext transport protocol (HTTP) model.

In 1993, the National Science Foundation created interNIC, which leads to the implementation of several Internet services, including formalized domain registration services by Network Solutions. Marc Andersen and the University of Illinois introduced Mosaic, the first graphical user interface for the World Wide Web. Andreessen would go on to start Netscape in order to further develop and commercialize the use of browsers. Compuserve and other subscription information services began offering businesses and home users access to the Internet through their information networks.

By 1995, large numbers of users were were joining the Internet community and business began thinking of ways to use the Internet to sell products and services. Since then, the evolution and growth of the Internet has been rapid.

It is important to note that the Internet and the World Wide Web are not really one and the same. The Internet refers to the massive network of networks, which is basically the hardware infrastructure. The World Wide Web is a combination of methods and protocols for accessing the information over the Internet.

Size

No one knows for certain because it is growing so rapidly. Estimates range from 12 billion web documents to over 20 billion. It could actually be much larger than that.

Internet for Military

The original concept developed by RAND was for a military communications network. It evolved into more of an academic study involving better ways to communicate information. For many years the Internet was only available to universities and research facilities. There was military funding and research from DARPA, which kept the project alive, so there was a strong military interest as the project evolved through the 1960s, 1970s and early 1980s. There are people who played important roles in the history of the Internet project who say it was not a military project, but that is a bit of a denial regarding where the concept originated and where much of the funding came from. It is safe to say that by the time the concept of the World Wide Web started to emerge, the Internet was no longer a military project.

FireFTP is a FireFox-only add-on, which means that it will not work with Internet Explorer. Installation is simple. Just go to the FireFTP download page and click on the Download button. Make sure you are using FireFox when you visit the page. The browser add-on will automatically install in a few seconds.

After installation, you can find the FTP utility in the FireFox Tools menu. Click on it and it will open in a browser window. Just like FileZilla, FireFTP allows you to set up and save FTP connection accounts and allows quick connects for occasional FTP connections that you do not want to store. The only difference is that an Account is saved, while a QuickConnect is not. When you start up FireFTP, you will see multiple panes. The panes on the left side represent the local directories on your PC. The panes on the right display remote directories on your web site after you have connected to your server. To set up an account, click on the Create an account... selection from the drop-down menu in the top left-hand corner. The Account Manager dialog box will display. From there you can enter the FTP information for your web site.

The Account Name will be displayed on the drop-down menu. The account information can be obtained from your hosting company. The Category is optional and can be used if you have a lot of FTP accounts that you want to set up in logical group sonthe menu. The Host is typically your domain name, *i.e.*, tech-evangelist.com. If you are using a dedicated IP address, you can use the IP address.

Some hosting companies require the use of special Host addresses:

- The Login is your username for the FTP account.
- The Password is the password for the FTP account.

To set up an account, you only need to enter the Account Name. Any information that you enter will be stored when you same the account. If you are using a laptop that could be stolen, it is a good idea to leave the password blank. FireFTP will prompt you for missing information when you try to connect to your server.

The Account Manager also has a Connection tab where you can set up any special connection information, such as a non-standard port number. You can also designate default directories for both the Local (your PC) and the Remote (the server) sides of the connection. This feature saves a lot of time because you do not have to search for files on your PC if you store them in a special location.

Files

Use the Browse button to the right of the path displayed just above the Local pane. You can select any location that you have access to on your network. Make sure that the proper directories are selected in both the left-hand pane (Local) and the right-hand-pane (Remote). Highlight the files your wish to transfer in the left window and click the right arrow in the center bar. You can also transfer files from the server to your PC. It is important to make sure that you have the proper directories selected.

To disconnect from your server, use the Disconnect button to the right of the drop-down menu. Once a connection is made, the Connect button turns into the Disconnect button. This is easy to set up, although the location for the change is not very intuitive. On the top right-hand side of the FireFTP window there is a Tools selection. Select Tools > Options. Check the box named Show Hidden Files. This is found in the Connection tab in the Account Manager dialog box. The easiest way to set this up is to make the connection to your server, select the proper directories for both the left-hand and right-hand panes, then select the Edit button next to the drop-down menu. Select the Connection tab. You should see both the Local and Remote directory paths displayed. Click on the Use Current button next to each, and then click OK. The next time you connect to your server, the default directories will be used. If you need a real nice, lightweight and simple FTP utility, it is worthwhile to take a look at FireFTP. I was pleasantly surprised with this one.

Web Server

A web proxy server, sometimes called a web proxy, can serve a variety of purposes. Many of the purposes for using a proxy server are legitimate, but others are suspicious and sometimes malicious. Using a simple definition, a proxy is basically a server that sits between a user and the destination server that the user is accessing. It is an intermediary or go-between that requests resources from the destination server on the user's behalf. On of the most common uses of a proxy server is for anonymous web surfing. When a user is connected through a proxy server, the IP address of the web proxy is used to communicate with the destination server and thus the IP address of the user is hidden. This can be used to allow a user access to systems where their access might normally be blocked, or because an IP address identifies a user's geographic location, it can be used to disguise the country where a user is located.

There are thousands of free proxy servers set up in countries around the world, and lists of these servers can easily be found on the web. When a user visits one of these sites, they enter the web address of the site they wish to visit. From that point on, the true IP address of the user is protected and the destination server only sees the IP address of the proxy server.

Students use web proxies to allow them to access web sites whose IP addresses are blocked by their school's server. For example, if a school has blocked all the known IP addresses for porn sites, students might still be able to access these sites through a proxy. Because of uses intended to circumvent normal security, lists of the IP addresses of known proxy servers are available, which can also be blocked by a company or school network. If you are concerned about protecting your identity while on the web, a proxy server can help with anonymity. But be aware that these servers can have a malicious intent and can be used to cache information that passes through them, which means that you never want to use one of these intermediaries to log into your bank account or any web site where exposing user IDs and passwords can be detrimental. This can be a path to identity theft.

There are other malicious uses for these servers. Many e-commerce sites capture the IP addresses of customers who purchase products. They are used to identify the geographic location of a customer to help prevent fraud. If a customer's street address is in Boston, but their IP address identifies their geographic location as Nigeria or China, the chances are more than good that a stolen credit card is being used. Many sophisticated online thieves use domestic drop points or freight forwarding companies where products are shipped and later forwarded to the thief. They have also learned to use web proxy servers to hide their location, or make it appear that they are located in a country not known for credit card fraud.

There are many legitimate uses for proxy servers. AOL has traditionally cached commonly visited web pages and altered JPG images by compressing them to speed up web access for their users. They can also be used by schools and companies to filter content, and by ISPs that restrict access to family friendly content. If you are going to use a proxy server, try to identify those that are legitimate. Many are, but some are not. As a general rule, never access sensitive sites that require a user ID and password. This includes your e-mail accounts.

Using PHP

A shopping cart is an essential component of every e-commerce site. Although there are literally hundreds of shopping carts to choose from, only a few stand out. PHP is the preferred shopping cart language because it is the most popular web development language on the Internet. When combined with the MySQL database, PHP is unbeatable for cost effectiveness. The selection of the best shopping cart for your web site is an issue that should not be taken lightly. The user-friendliness built into your shopping cart can be a major factor when it comes to retaining visitors and converting them into customers. Numerous studies have shown that the checkout process associated with a poorly designed cart can be a major factor affecting shopping cart abandonment during the checkout process. The following is brief list of the most popular shopping carts that utilize both PHP and MySQL. Both PHP and MySQL are free (no licensing issues) and are offered with almost all non-Microsoft web site hosting. This gives this combination a low-cost advantage, both for the cost of obtaining, modifying and maintaining your shopping cart system. Indeed, many of the most popular shopping carts using PHP and MySQL are free and most have a wide range of support available through numerous forums.

All of these shopping cart options are complete e-commerce storefronts, complete with category pages, product pages, checkout processes and much more. All have a wide range of mods available that add functionality to the e-commerce sites.

Os Commerce

First and foremost is os Commerce, the biggest player in open source shopping carts. osCommerce has been around for more than six years and includes a large community of users and supporters. osCommerce is a complete e-commerce storefront. The basic osCommerce download is very functional and there is a wide selection of plugin mods that add functionality and features. More than 100 templates are available that change the look and feel of a storefront.

When considering shopping carts, CRE Loaded has been referred to as osCommerce on steroids. CRE Loaded is built on osCommerce. Different versions include large collections of already-installed mods so that you have a highly functional and heavily equipped version of of osCommerce as soon as it is installed. There are small one-time fees associated with purchasing a copy of CRE Loaded, but the costs are minimal. Zen Cart is a derivative of osCommerce. The group that focuses on this cart started with a version of osCommerce a few years ago and used it as a basis for building their own open source shopping cart. it is therefore similar to osCommerce in some ways, but has its own user and developer community. Support is provided through a very active forum where members help each other to solve issues related to e-commerce and the use of Zen Cart. The best part is that it is a very stable cart with loads of built-in functionality and it is free.

This is another contender when it comes to popular and very functional shopping carts. Cube Cart has both a free and low-cost version available. They are both the same, but it will cost you $89.95 if you wish to remove the Cube Cart copyright notice and link from the page footers. Sometimes this makes sense if you are in a competitive market and you do not want your competitors to know which shopping cart you are using.

Observations

Any of the choices listed above are good choices for an e-commerce web site. One thing to note is that most of the systems have all of the functionality turned on when you install the cart. This tends to degrade speed performance due to numerous unnecessary database lookups. After installing one of these shopping cart systems, you need to become familiar with the administration area and then should turn off every feature that you really do not need. Most of these systems have manuals available, either in hardcopy or downloadable format. Your learning curve will likely be shortened if obtain the manual and go through it in great detail. Which one did we like best? We decided to go with Zen Cart due to the very active user forum and the superior support we see from the development staff. You should, of course, make your own decision. You cannot go wrong with any of these PHP shopping cart choices.

Troubleshooting

Outbound e-mail uses a process called SMTP (simple mail transport protocol). SMTP is the standard for e-mail transmissions across the Internet. SMTP is generally used to send messages from a mail client (Thunderbird) to a mail server. The e-mail is stored on the server until retrieved by another e-mail client using a POP3 or IMAP retrieval protocol. Outbound e-mail problems can be frustrating to deal with because you rarely see an indicator as to what might be causing the problem. However, there are fewer factors involved when setting up the SMTP section in Thunderbird, so isolating the problem can be much simpler. Unlike setting up an e-mail retrieval for inbound e-mail, with SMTP, a single SMTP setup can be used to cover all of your e-mail addresses, regardless of how many different ISPs are involved.

Most SMTP mail servers are set up to allow e-mail accounts from other ISPs to be run through them. Thunderbird identifies the first SMTP account that you set up as being the default account. That is because in most cases it can be shared by all of your e-mail accounts.

Most of the configuration issues with sending e-mail can be resolved in the Server Settings section in Thunderbird.

- Select Account Settings from the Tools menu.
- Scroll down to the bottom of the list if e-mail accounts.
- Select Outgoing Server (SMTP).

When you are sure that this information is correct, click the OK button, shut down Thunderbird and then start it up again to make sure that any new settings are loaded. If you have multiple e-mail accounts and you know that inbound e-mail is working, try sending a test message to another of your e-mail addresses. If it still does not work or you see an error, make sure that you Internet access is active. If you can access the Internet using your browser, you should be able to send e-mail messages once the SMTP information is correct. If you are still having trouble connecting to the SMTP server, Thunderbird will usually display a message, but depending on the type of error, sometimes it does not. Most of the time the User Name is just the account name, but some configurations may require a complete e-mail address. If one way does not work, try the other.

If your SMTP server requires a secure connection, you may have to check the SSL box. I sometimes use the SMTP server at my AT&T

account. AT&T connections usually require that you check the SSL box. If you are having connections problems, post your issues in the Comments below. Try to give us as much information as you can and we will see if we can help you to get connected.

Incoming E-Mails

Most e-mail addresses today use the POP3 (Post Office Protocol 3) Internet protocol for receiving e-mail. With POP3 your e-mail is held on a mail server. You can download your mail to an e-mail client. Typically, the e-mail is deleted from the server when it is downloaded, but it can be set up to retain the e-mail messages.

An alternative Internet mail protocol is IMAP (Internet Message Access Protocol). IMAP provides some enhanced capabilities and is useful for dealing with large volumes of mail that may need to be organized. Most hosing companies today use POP3. If you are not sure about your situation, ask the ISP or hosting company. The e-mail documentation they provide will usually tell you which protocol you are working with.

Most of the configuration issues with receiving e-mail can be resolved in the Server Settings section for each e-mail address in Thunderbird.

Each e-mail address is configured separately:

- Select Account Settings from the Tools menu.
- Select Server Settings under the e-mail address you wish to troubleshoot.

The first thing to check is the Server Name. This identifies your mail server. You will need to get this information from your hosting company or ISP. Most mail servers are identified by using a subdomain preceding the domain name, such as mail.tech-evangelist.com. Some use other subdomain names, such as ipostoffice.tech-evangelist.com. Once again, the only way to know is to find out how your mail server is configured. Next, check the User Name. Sometimes this is the entire e-mail address and sometimes is is just the account name, which is the portion of the e-mail address to the left of the @ sign. The third item to test is the Security Settings. Most mail servers do not use a secure connection, so Never would be checked. I've seen a few that use TLS (Transport Layer Security), which is a type of secure connection. If Never didn't work, try TLS, if available.

There are also a few e-mail servers that use SSL (Secure Sockets Layer) for secure connections. Secure connections encypt the messages as they are downloaded, which prevents others from intercepting your messages. I have an AT&T e-mail account that requires SSL. When you click the SSL radio button, the port number will change. You can also try checking the Use Secure Authentication box, but thus far I have never had to do that when setting up an e-mail account.

There is a very small possibility that your ISP or hosting company is using a non-standard port. Thunderbird does allow you to change the port, but you should never do that unless instructed to do so by your ISP or hosting company. If the port is wrong, you will not make the proper server connection. If none of this works, make sure that you do indeed have an active Internet connection. If your browser can access the Internet, you should be able to retrieve your e-mail.

By now your problems receiving e-mail should have been resolved. Most of the time when I've run into issues setting up new e-mail accounts in Thunderbird, the issue was an incorrect User Name or Security Setting. Remember to save the settings each time you make a change by clicking the OK button. It is also best if you shut down and restart Thunderbird to make sure that your changes are reloaded. After that, test it again.

Creating E-Mail Signature

A signature in an e-mail is kind of like a footer on a web page. It is generally used to convey contact information, legal notices and other repetitive information. This is something that you may not want to re-type every time you send an e-mail message. Thunderbird allows you to create as many signatures as you wish. You can assign a different signature to each individual e-mail address.

The first thing you need to know is that you must create the signature file in a pure text editor, such as Microsoft's Notepad. Do not use Word or any type of text editor that may embed codes in the text. You can save these files anywhere on your PC, but it makes sense to store them somewhere where they will not get separated from the rest of your Thunderbird files. In a previous tutorial, I changed the file location for the Thunderbird files in order to make it easier to back up these files.

I simply created a new folder called "signatures" in the folder with the rest of the Thunderbird files. Like I said, you can store them anywhere, but when you do it this easy, the signatures stay with your Thunderbird files if you need to move them to another PC or if you have to restore a PC from your backup files.

Next, you will need to assign a signature to each e-mail address. From the Tools menu, select Account Settings... Highlight an e-mail address on the left-hand column. On the right, check the box that says "Attach this signature:" and then click the Choose button to navigate to your signatures folder. Select the file that you want to attach. Click the OK button to save your settings. You will need to repeat this process for each e-mail address where you wish to add a signature. If you add a URL to the signature file, most e-mail clients will turn it into a hyperlink. But if you need to spruce it up of if you wish to use HTML in a signature instead of plain text, here is what you need to do. Create a text file in the same manner as described above. When you save the file, save it with a.html extension, rather than a.txt extension. Thunderbird will then recognize it as an HTML file. You can then use coloured fonts, hyperlinks or just about any type of CSS, as long as you use inline CSS styles.

When you are using CSS in e-mail messages, make sure that you fully specify the URLs to images and store the images on your server. Don't try to embed images or send them with the message.

It won't work well with many browsers.One more thing that has to be mentioned about HTML in signatures is that the message must be sent in HTML mode for the HTML and CSS to work. When you send any message that contains HTML, Thunderbird will ask you if you want to send it in in Plain Text. You do have to click on the Send in HTML Only or Send in Plain Text and HTML for the HTML and CSS to render properly.

Spam and Junk Mail

No one want to receive spam, but today it is almost inevitable that most e-mail addresses will eventually end up on spam lists. I have always protected my e-mail addresses, but recently ended up receiving more than 100 spam messages per day on my most important e-mail address used only by my business clients.

It turns out that a client thought she was doing me a favour when she posted it on a forum, along with a recommendation. As part of my battle to eliminate spam, I found that Thunderbird contains some amazingly powerful spam and junk mail controls that are very effective at filtering unwanted e-mail. This tutorial shows you how to turn these filters on and configure them to meet your needs.

There are actually several methods for filtering out spam using Thunderbird's tools. I used all of them in conjunction and found that Thunderbird effectively filtered out almost every undesirable message.

It is also amazingly accurate and only very rarely grabs a legitimate message and identifies it as spam.

Pre-Filtering Spam Messages

The first level for setting up spam filtering is to do it at the server level or on your personal computer. If you have a web site, your hosting company might already offer spam filtering software, such as Spam Assassin or Spam Pal. This step is not absolutely necessary, but it does make the process of identifying and filtering spam more effective.

SpamAssassin is commonly found on Linux or Unix Apache servers. SpamAssassin checks messages coming through a hosting service's e-mail system. It does not actually filter the spam out, but rather uses heuristic algorithms that analyse a message. It then assigns a spam score. Thunderbird can use this score to determine if a message is likely to be spam. If your web site runs on Unix or Linux and uses a control panel such as Pleske or cPanel, check to see if SpamAssassin is included. You will find it under Mail, SpamAssassin. If the message near the top indicates that SpamAssassin is disabled, click the box labeled Enable SpamAssassin. Do not click the box labeled Enable Spam Box.

SpamPal is a different type of filtering system that installs on your PC. SpamPal uses spam block lists (lists of IP addresses used by spammers) to identify spam messages. I have not tried it, but if your hosting company does not use SpamAssassin, or if you do not have a web site (and therefore do not use a hosting company), you might want to try installing it. You will find it atspampal.org.

Configuring Junk Mail

The first step is to configure Junk Settings:

- Select Account Settings from the Tools menu. Thunderbird allows you to configure each e-mail address separately.
- Select Junk Settings under one of your e-mail addresses.
- Check all of the boxes in the Junk Settings display. If you don't use SpamAssassin or SpamPal, skip that checkbox. If you do use one of these tools, make sure the box is checked and the proper filter selected.
- Checking "Move new junk messages to Junk folder" will create a Junk folder under the selected e-mail address. You can set the nuber of days before spam is deleted from the junk folder. It is a good idea to move the messages to this folder so that you can periodically review it for messages that may not be spam.

Configuring Privacy Settings

- Select Tools, Options.
- Select the Privacy button.
- Select the Junk tab.
- Choose how you want spam to be handled when you manually select it when reviewing your e-mails. When you are viewing messages and you identify a spam message, you can click on the column to the left of the date to send the message to the Junk folder. This also helps the system identify spam messages.

Configuring Message Filters

When all else fails to identify spam, this is the way to do it. Message filters will search new messages for any keywords you designate. This section is pretty easy to figure out. Once again, you have to set up individual message filters for each e-mail address.

- Highlight an e-mail address in the main screen.
- Select Tools, Message Filters from the top menu.
- Click on the New button.
- Name your message filter at the top.
- Select the e-mail element that you wish to filter. Most spam keywords are going to be found in the Subject or Body.
- Select Contains.
- Enter the keyword or phrase you wish to use as a filter.
- Near the bottom of the displayed dialogue box, select the location to move any identified spam messages. The Junk folder for the current e-mail address is the most logical place for these messages so that you can review them.
- Click OK.

That's about all there is to configuring Thunderbird to filter out spam messages. Although the configuration sections are a bit scattered throughout Thunderbird, once you know where to find them they are easy to use.

View Source

If you are an avid Internet user and use Internet Explorer as your browser, you have probably noticed that there are times when the View Source function in Internet Explorer occasionally stops working. For a Web developer or designer who likes to view the HTML code for a Web site, this can be frustrating. Fortunately, both the explanation to the problem and the solution are usually simple.

The root of the problem is that the browser cache has filled because of all the Web page code and images stored by the browser. The solution is to clear these temporary files and start out with a fresh cache. To do this,

- Select Tools on the top menu in Internet Explorer
- Select Internet Options...
- Click Delete Files in the Temporary Internet Files section

It may take several minutes to delete all the files in the cache, so don't panic if it appears like your PC has locked up. As soon as the hour glass once again turns into a mouse pointer, you can shut down Internet Explorer and start it up again. The View Source feature should now work once again.

If this does not fix the problem, you may have a more serious issue, such as a corrupted or deleted Notepad.exe programme file or problems related to the temporary files directory in Windows. Those problems are beyond the scope of this article. Other issues that have been reported that sometimes disable View Source include having a shortcut to Notepad on your desktop or having a shortcut named Notepad on your desktop. Renaming the shortcut should resolve those issues.

Most of the time, the problem tends to be related to the cache, so if you spend a lot of time on the Internet, it might be a good idea to clear the cache once a week.

Mozilla Thunderbird Tips and FAQs

Mozilla Thunderbird is, in my opinion, the absolute best free e-mail programme available. It doesn't have the vulnerabilities found with Outlook, primarily due to the fact that it is not an "attack target" like Microsoft products. Even when vulnerabilities are found, they are quickly patched by the community of developers that support Thunderbird. We've put together a list of tips to make your use of the Thunderbird e-mail application more enjoyable.

If you have a specific question regarding how to do something with Thunderbird, post a comment below and if we have a solution we will include the answer on either this tips page or in a separate tutorial.

Where do I Find the Download for Thunderbird

You can find the most current version at the Mozilla Thunderbird download page. Installation is pretty simple. If you are using Outlook, Eudora or another e-mail client, you can easily import your address book and messages during the installation.

Including Previous Message Text

We found this to be one of the annoying default features with Thunderbird. Unless you change the settings, when you forward a message with Thunderbird it will include the forwarded message as an attachment. I personally prefer that the forwarded text be included in the body of the message that I send.

This is easy to change:

- Click on the Tools menu
- Select Options
- Select Composition
- Select the General tab
- *At the top change Forward messages*: "As Attachment" to "Inline"
- Click OK

Your forwarded messages should now include the text of the forwarded message in the body of the e-mail just below any message text that you add. If you selected the default settings during the installation, Thunderbird will automatically download updates and let you know when they are ready for installation. You will find the settings for this under Tools, Options, Advanced and then select the Update tab. Check out our tutorial about setting up Thunderbird mailing lists.

This is really simple to do, but like many things with open source software, the solution is not always intuitive. This is an area where Thunderbird excels. We did a separate post on that called Thunderbird spam and junk mail filters. We tried several different methods to eliminate spam and this was the only system that eliminated it almost completely and at zero cost. We were very impressed.

String E-mail messages

They are hidden, but you can find them using the tips we offer in Thunderbird file location. This is very useful information if you wish to move the location to a Windows folder that can be easily backed up.

Changing Passwords

Ah-h-h, grasshopper. Passwords used to access your e-mail accounts are easy to change if you follow the instructions we offer in Changing Thunderbird Passwords

Making Mozilla Thunderbird

When you set up Thunderbird as the default e-mail client, it will automatically open whenever you click on an e-mail address on a web page that uses the HTML mailto: protocol. To configure Thunderbird as the default e-mail client, simply Open the Tools menu, select Options and click on the General icon. Under system Defaults, check the box that says, "Always check to see if Thunderbird is the default e-mail client on startup." Click OK to save the setting. From now on, whenever the PC is booted it will check to see if Thunderbird is the default e-mail client. If one of the Windows programmes tries to change the setting to Outlook or if another programme changes it to something else, a message will be displayed that will allow you to change it back. Most people are not aware of this, but when you delete an e-mail message with most e-mail clients, you do not actually delete the message. The message is simple flagged so that it does not appear in the e-mail list. Thunderbird never actually deletes a message unless you take some extra steps. First, select File, Empty Trash. Second, select File, Compact Folders. This deletes all of the flagged messages in the data files. This issue keeps coming up. It looks like you can use Thunderbird with a Yahoo account if you subscribe to Yahoo Mail Plus. The standard freebie Yahoo account does not offer POP3 access. See the following articles. You will find all of the proper configuration information posted there.

- Can I use Thunderbird to read and send my Yahoo! Mail?
- Can I POP my mail into a different e-mail client (like Thunderbird)?
- Yahoo outgoing e-mail problems with Thunderbird.

Preventing Thunderbird

This was a feature that was turned on by default in older versions of Mozilla Thunderbird. I think the default is turned off in newer versions. It is something that is not only annoying, but also presents security issues.

Turning the embedding feature off is easy. Just select the View menu. Make sure that Display Attachments Inline is unchecked. Changing Thunderbird's behaviour with outgoing e-mail attachments is easy to do, but is buried in the Config Editor. Make sure that you are using a current version of Thunderbird.

The location of the Config Editor has changed several times with different versions. Select the Tools menu, then Options, select the Advanced button, then the General tab and click on the Config Editor button. Scroll down until you find mail. content_disposition_type. If the value is not set to 1, double-click on that line and an Enter Integer Value dialog box will pop up. Change the value to 1 and click OK. Shut down Thunderbird and restart it for the change to take effect.

Programmer

When a non-technically inclined web site owner speaks with an experienced programmer, technical jargon frequently enters the conversation. Some programmers do that intentionally, just to show off their expertise. For others, it is just the way they talk because they are immersed in a technical working environment. If you get confused talking with IT people, you may find this jargon guide to be handy.

The technical language of programmers can sometimes be very difficult to understand. It is loaded with shortcuts, euphemisms and analogies. Frequently, the full name or technical term for the many special symbols or characters used in programming is just too cumbersome or boring to use in everyday technical conversations. Unique names or monikers for programming symbols has evolved. Most are intended to be humorous and help to liven up the communications between programmers.

The following are a few of the most common jargon and techno-terms used by professional programmers in everyday conversations amongst themselves. Step outside of your box for a few moments and compare the special characters to the jargon terms and you can easily envision the evolution of the terms.

- *. (dot)*: A dot is just a period in a file name or domain name, as in google-dot-com.
- ** (star, splat or spider)*: An asterisk is a wild card symbol used with servers and computer programmes.
- # (hash, pound, crunch, sharp)–Most commonly referred to as a hash, this symbol is commonly used to designate comments in many programming languages.
- *! (bang, pling, shriek, not)*: The exclamation point serves different purposes in different programming languages, but it is most commonly used to reverse a Boolean evaluation. In other words, if a statement is true, it reverses it to false.
- *#! (hash-bang or sh-bang)*: When used as the first two characters on the first line of a programming script, a hash-bang causes Unix and Linux operating systems to execute that script using the interpreter specified by the rest of that line.
- *‘ (tick):* This is a single quote commonly used to surround a string of characters.
- *“ (snakebite, rabbit ears)*: A standard double quote. This jargon goes way back to the 1980s, so if you hear this, you are talking to an old-timer.

- *$ (buck, bling, cash)*: Used to designate a string variable in many programming languages.
- *() (left paren, right paren, open, close)*: Commonly used to group portions of algorithms or comparison statements. Algorithms and evaluations of statements always start with the inner-most set of parens in all programming languages.
- */ (wack, slash, stroke)*: A common slash character, most commonly used in directory paths on servers or to define a mathematical division.
- *\ (backslash, backwack, hack)*: A common blackslash used in some networking paths and also as a common designation for a escape character, which means that the next character in line should be taken literally and not executed.
- *{} (left curly, right curly, hitchcocks*: Think Alfred Hitchcock silhouette, left banana, right banana)–Curly braces commonly used to group multiple programming statements.
- *~ (squiggle)*: A tilde character.
- *| (bar, pipe)*: A vertical bar character.
- *^ (hat)*: A caret character.
- *% (mod, grapes)*: Commonly used as a mathematical modulo character in programming languages.
- *& (amp, amper)*: An ampersand symbol.

W3C

You may find it interesting to learn how to eliminate every error message the W3C validator produces. It can be an educational process even if you are a very experienced developer. Although you may think you know a lot about HTML or XHTML, you will likely find that you do not know as much as you thought you knew about the W3C's coding standards. Before you use the validator, a page needs to have a Doctype declaration. The Doctype declaration, or DTD, needs to be the very first line of code on a Web page. The DTD sets the standard to be used in the validation. It also serves another purpose. Many modern browsers contain multiple rendering engines. If you do not set a standard to be used for rendering, the page may not look consistent across different browsers.

Note that all valid DTDs from HTML 4.01 on contain a link to a standards document. If your web page editor already adds a DTD with every Web page, it may not be valid if it does not contain a URL to the actual standards document. Using the W3C validator is very easy. Just cut-and-paste or enter a complete URL to the Web page you wish to test. You may find some of the validation messages to be a bit cryptic, but the W3C provides links with most error messages that lead to additional information. There are some common page elements that always produce errors or warnings, even though they do not negatively affect either browsers or search engine spiders. Microsoft-specific entity codes, such as copyright and other symbols may not be code compliant. The validator handles these errors well and typically offers the code compliant equivalent.

Body tag attributes such as leftmargin, right margin (Internet Explorer), marginheight, marginwidth (Netscape) are browser extensions that are not part of the official coding standards. These types of issues will not typically produce negative effects with either browsers or spiders, but do show up as errors because they do not comply with the standard. One type of nuisance error is a missing alt attribute in image tags.

Current standards do require the use of the alt attribute in order to push designers and developers to use this attribute because it is beneficial for Internet users with vision impairments who utilize special browsers that describe images though the use of alt attributes. It may not make sense, however, to include text with every alt attribute. The code compliant workaround is to simply include an empty string value (alt="") for lines and other page elements that need no explanation.

It can sometimes be very beneficial to eliminate every error message displayed, and to assure complete and 100% compliance with the DTD standard you chose. Once a page has been validated as compliant, the W3C offers a cut-and-paste snippet of code that can be added to your Web pages as both a "gold star" certification of compliance and as a link to re-validate the code when you make changes.

Avoiding Spam

Despite anti-spam laws, the amount of spam e-mail messages I receive has grown enormously. You probably feel the same way. The problem with spam laws is that someone has to enforce them, but the e-mail spam problem is so enormous that only the largest and most aggressive USA spammers get prosecuted. Foreign spammers are not subject to USA laws and are much harder to identify. There are, however, some fundamental things that you can do to lessen the amount of spam that you receive, particularly if you have a business.

First and foremost, the best way to prevent your e-mail address from finding its way to a spam list is to never, under any circumstances, publish the e-mail address on a web page. Why? Because most e-mail ends up on spam lists because one of the hundreds of spam bots–also called e-mail harvesters–found it on a web site and automatically added it to a spam list. If you need to publish an e-mail address, create an image file for the e-mail address and display that. Spambots cannot read text in images, so they will not find the e-mail address if you display it in an image.

Second, never, ever click on a spam message or any links in a spam message. This includes links that assure that the message will be removed from the spammer's database. While this may work for legitimate sources of e-mail, I am willing to bet that few spammers really care about the laws, especially if they are in a foreign country. When you click on any link, you are very likely just verifying that your e-mail address is valid, which pretty much guarantees that it will be added to hundreds of spam lists.

Third, much of the spam that we receive comes from spammer programmes that randomly generate e-mail account names using the domain names of popular ISPs and then send an e-mail to that address. If it bounces back, they know the e-mail address is not valid. If it does not bounce back, it may be a valid e-mail address.

Many spammers use a library of account names that are attached to other libraries of domain names. The account name if the portion of the e-mail address found to the left of the @ symbol. The domain is the part to the right.

Due to the fact that we run several web-based businesses, we have noticed a pattern of account names that many spammers use to see if an e-mail address is valid. The account names are attached to lists of domain names that can easily be found on the web and are used to generate spam for commonly used e-mail addresses.

Internet Technology

The Internet has become critical to everyday life in domains as diverse as education, health, defence, commerce, travel and entertainment. The Internet was not designed for its current level of usage, and there is a need for simple constructs to allow the network to do better in terms of security, mobility and quality of service, among other things. A number of global and local programmes of research (in the US, EU and further afield, eg Asia) are looking at future network architectures and building testbeds to evaluate new protocols and systems based on these new ideas. The most notable, in the US are the NSF Find programme and the GENI project to build infrastructure. In

Europe we find a number of projects, with some interesting high-level thinking (eg the Eiffel project) and a set of testbed initiatives under the FIRE Programme. While some researchers think there is a need for a'clean slate' design of the network of the future, much successful work is evolutionary rather than revolutionary, and this can be seen in the articles in this issue. Future core networks will leverage IP over simple, super-fast optical core networks. A major trend is that of virtualization, featuring the construction of optimized virtual networks that answer the needs of a collection of users or applications. A wide variety of wired and wireless access networks are available or being developed: FTTH/FTTO in the wired domain, and 3G/CDMA, LTE, WiFi, WiMax and Satellite in the wireless setting.

Internet Modelling

It is quite important to comprehend how the complex systems that we build actually operate. This requires basic research on network modelling and simulation, eg, in order to derive the fundamental laws on network dynamics and control or to evaluate the ultimate capacity of self-organized wireless networks. It also requires advances in network measurements: traffic statistics, Internet probing and measurement, network inference and detection of anomalies and attacks.

Internet Algorithms

The future of the Internet will require a wide range of computer science tools: verification; distributed algorithms (eg for consensus, and election and epidemic diffusion); resource management algorithms: resource allocation and scheduling; database algorithms: content storage, update and retrieval, content replication and consistency; search engines and the semantic Web.

A key cost in networking is the operations and management overheads. The scalability of the Internet is well known, but as distributed applications proliferate, a more autonomic approach is increasingly required. Spontaneous and self-organized networks emerge both in the wireless setting (eg with Wifi meshes or infrastructure-less wireless networks [MANETS]) and in the wired network setting (eg in peer to peer). The Internet started by connecting computers and users to information, and then went on to connect users to each other with audio, video, games and social networking tools. Now the rapid evolution of pervasive, embedded networked devices means that we connect various types of devices to each other. There are far more computers in the world embedded in everyday objects (cars, domestic appliances etc) than there are on desktops.

The interconnection of the Internet with the physical world through sensors and agents and the tagging of industrial production by RFIDs will lead to new traffic and architecture challenges, with possibly hundreds of billions of new devices that will collect information and will have to be upgraded and managed remotely and conveniently. This will require new paradigms for routing, search, naming, maintenance, data survival etc.

The Internet of things extends to cars and other vehicles. Here the key potential is in safety: ABS in cars could communicate road surface conditions to following vehicles, setting speed and braking reactions sooner rather than too late. Traffic management and pollution sensing on cars can also make use of networking between cars, and from cars to roadside infrastructure. Tracking goods in transit on the road would allow logistics companies to optimize their freight operations, saving time and energy and perishable goods.

There is an important diversification of the nature of the content transported across the Internet: initially it was files, then real-time games, video, telephony and whiteboards, and now TV, video on demand etc. There has also been diversification of the localization of contents: with each user potentially a content producer (peer-to-peer applications). New interactions with data are appearing, as in Web 2.0 or the semantic Web, and applications continue to evolve and require new systems, measurements and management tools. Since its beginnings in 1992, the World-Wide Web has offered remote transactions for goods and services. Recently we have seen a rapid growth in the number of attacks on identity, since acquiring such information allows miscreants to commit fraud that is hard to detect.

Estimates vary, but the Internet and all its services consume something on the order of 4% of the energy in the developed world. Only simple measures are required to improve this by a factor of two. Furthermore, the Internet, as we have discussed above, can be used to monitor and control external devices (things, vehicles, services) and significantly reduce their unnecessary power consumption. Figures as high as 30% have been quoted for possible national savings of energy, if unused devices in all homes could be remotely turned off. The investment necessary to achieve this is relatively low, with what seems like a very big potential return.

The Internet Services

We are moving from a Web of documents to a Web of services and Web of knowledge. This has triggered an explosion of new applications such as SecondLife, FaceBook and LinkedIn. And there is more to come, with augmented reality, virtual worlds, real-time games and telepresence. New concepts such as service orchestrations are also emerging, and with them, new business models.

Setting The Context

- The invited article 'Starting the Debate: Agreeing on Disagreements' by Dirk Trossen, B.T., U.K., is a meta-discussion note, on how to trigger debate. As such, it is quite interesting as a methodology for encouraging architectural thinking in the long term.
- 'The Internet Engineering Task Force and The Future of Internet' by Emmanuel Baccelli, Thomas H. Clausen and Philippe Jacquet, INRIA Saclay, France, provides a useful reminder of the processes by which standards are created, and an analysis of how future research might be delivered in the short to medium term.

Architecture

- 'Network Virtualization, a Perspective' by Anja Feldman, Mario Kind, Olaf Maennel, Georg Shaffrath and Christoph Werle: Virtualization, briefly defined above, can be seen as a way to overcome the ossification of the Internet, namely the general idea that one should not change a system or architecture that works well. The technical challenges and the business models associated with overlay networks and more generally with virtualization are thoroughly discussed in this invited paper.
- '*Y-COMM*: A New Architecture for Mobile Heterogeneous Communications' by Glenford Mapp suggests that the symmetric architecture we are used to in the Internet is perhaps not well adapted to wireless, and proposes an alternative, useful for promoting discussion (cf the EU ANA Project).

Modelling, Measurement and Management

- 'Epidemic Information Dissemination' by Laurent Massoulie (Invited paper): Epidemic algorithms became popular at the turn of the century with Peer-to-Peer applications. They are now being investigated by major companies designing efficient distributed live streaming mechanisms over the Internet. This beautiful line of research combines the design of innovative algorithms and the modelling of their execution over a large network, which allows one to prove their quasi-optimality.
- Understanding traffic entails delivering information efficiently to humans so that they can decide on actions to take. The article

'iMyNetScope- a Platform for Network Traffic Visualization and Analysis' by Pavel Minarik discusses one approach to this, and presents one particularly interesting topic for today - the element of Cyber Defence.

- 'Unified Access to Internet Measurement Data' by Felix Strohmeier, Martin Nilsson and Demetres Antoniades: like the previous article, this touches on a timely topic - making sure we can all (as researchers or operations) access data easily (see also www.crawdad.org for an example of traffic archives in a standard form). Since the Internet is a federation of a vast number of networks (300 ISPs in the UK alone), we need clear and standard methods for exchanging data about operations. This applies equally to research networks such as those in FIRE.
- 'Privacy Aware Network Monitoring' by Peter Dorfinger, Carsten Schmoll and Felix Strohmeier describes a problem relevant to the research and operations community. Much work on the future Internet requires monitoring of research, but at the same time, real users require privacy. This article reminds us how important and difficult it is to reconcile these two aspects. Indeed, in gathering data about network use there are legal requirements relating to intercept laws and user privacy that must be met.
- 'Breaking the Weakest Link: Becoming a Trusted Authority on the Internet' by Marc Stevens is a reminder of the danger of security failures. The fact that even the basic technology of security is sometimes flawed is interesting. Systems are never perfect, and this is never more true than for security: we can never prove a system is secure, but only discover (eventually) when it no longer is!

Correctness and Robustness (and Self-Managed to Some Extent)

- '*ResiliNets*: Resilient and Survivable Networks' by David Hutchison and James P.G. Sterbenz: As discussed above, we need secure networks that are efficient and well managed. We also depend on these networks, and since individual components are never 100% immune from faults and failures, we need techniques to make the overall system more resilient than its components.
- 'Standardized Testware for Internet-Based Telecommunication Services' by Bostjan Pintar, Axel Rennoch, Peter Schmitting and Stephan Schulz: As Internet protocols grow in number and complexity each year, we need assurance that systems will behave correctly. This article outlines industrial solutions on

the correct operation of the software systems (protocols) that make the Internet and its telecommunication infrastructure work.

- 'Network Description Tools and Standards' by Freek Dijkstra, Jeroen van der Ham and Ronald van der Pol: Setting up high-speed network connections - light paths - is still a manual effort taking two to three weeks. The network engineers need a clear picture of the network topology in order to plan and configure light paths. The Network Markup Language (NML) standardizes network topology and state information. The University of Amsterdam and SARA Computing and Networking Services in the Netherlands are contributing to this effort, with the ultimate goal being to automatically set up and manage light paths.

Novel Applications, Especially Multimedia

- 'Delay-Tolerant Bulk Internet Transfers' by Nikolaos Laoutaris and Pablo Rodriguez (invited paper): Using disruption-tolerant networking (DTN) techniques, but on the Internet rather than on challenged networks, produces very interesting results in terms of capacity. Throughput can be significantly increased for some applications, and there are some very well-motivated examples, such as backing up information between large data centres, using spare capacity in a smart and autonomic way.
- 'Resource Management for IPTV Distribution' by Henrik Abrahamsson and Per Kreuger: IPTV on AT&T and Telefonica's networks now has in excess of 12 million users, each with hundreds of channels of broadcast quality content. This makes for non-trivial deployments,– and the importance of resource management for the various different technologies is unquestionable.
- '*VISTO*: Visual Storyboard for Web Video Browsing, Searching, and Indexing' by Marco Pellegrini describes exciting work on future Internet killer applications.
- 'The New Role of Humans in the Future Internet' by Daniel Schall and Schahram Dustdar: Current service-oriented architectures typically orchestrate Web services. This article addresses architectures that combine Web and human services, where the latter are not limited to service consumption but may also undertake service production. The main emphasis is on the interfaces required in such architectures.

In The Home (Internet of Things, And Energy!)

- 'Orchestrating your Surroundings' by Pau Giner, Carlos Cetina, Joan Fons and Vicente Pelechano, UP Valencia, Spain, is an example of the 'Internet of Things' being at the centre of future Internet visions.
- 'An Autonomic Home Networking Infrastructure' by Thomas Luckenbach, Mario Schuster and Marc-Oliver Pahl: in line with the previous article but articulated with a vision of autonomic computing, this is on the important topic of home net automation. Home users are non-technical: they really do not want to become network managers, operators or system administrators. It is thus essential to make home networks 'plug and play'.
- 'Load-Balancing Energy-Usage of Household Appliances' by Lennart E. Fahlen: Energy is a highly relevant topic, as described above, and using the Internet to control consumer electronics in the household is a promising way to reduce unnecessary consumption.

On the Road, Vehicular

- 'The Internet of Vehicles or The Second Generation of Telematic Services' by Markus Miche and Thomas Michael Bohnert: Like the Internet of Things, vehicular networking is just taking off. There are car-to-car and car-to-infrastructure communications standards as well as integration of satellite navigation, traffic monitoring, tracking, and cellular map systems.
- 'Web-Enabled Tracking Operations in Distributed Supply-Chains' by Zsolt Kemény and Elisabeth Ilie-Zudor: Using the Web to track logistical information about goods in the real world is already a huge business. This is a very useful article on the topic, which also relates to the previous article.

Awareness, Identity And Society

- 'Knowledge-Based Collaboration Patterns in Future Internet Enterprise Systems' by Gregoris Mentzas and Keith Popplewell is supporting virtual community patterns, a bit like eScience community work.
- 'Experiences from the Public Safety Arena for the Future Internet' by Socrates Varakliotis, Peter Kirstein and Steve Hailes: Since the Boxing Day tsunami and Hurricane Katrina, it has become clear that the Internet might be better used in support of communications during disasters. This is a useful note on work in the area.

- 'With Joint Forces - Establishing Situation Awareness for the Future Internet' by Tanja Zseby and Thomas Hirsch: Like Mentzas' article, we can build systems to track patterns of use. This can be used to support collaboration or to detect misbehaviour. Increasingly, though, agencies are beginning to monitor social networks for this purpose.
- 'Social Networking for a Pervasive Future Internet: the SOCIALNETS Project' by Stuart M. Allen, Marco Conti, Andrea Passarella and Roger M. Whitaker: A timely article on an increasingly important topic. This outlines several new aspects of social networking that our readership should know about.
- 'Owner-Centric Networking: A New Architecture for a Pollution-Free Internet' by Claude Castelluccia and Mohamed Ali Kaafar proposes a new Owner-Centric Networking architecture that provides individuals control over their contents. This architecture would considerably improve privacy on the Internet by limiting data pollution.
- 'Semantic Web for e-Commerce' by Bernd Gruber: The Web is the basis for most new applications (Web mail, photo sharing, social nets, maps) and is the basis for commerce, which drives real economies.

Wireless, Embedded

- 'Management of Future Mesh-Based Radio Access Networks' by Vangelis Angelakis, Vasilios Siris and Apostolos Traganitis: Community mesh wireless nets are becoming pretty widespread in some areas as an alternative to expensive cellular data services. Automating management is important; related work at Intel and Microsoft in the last couple of years has covered some aspects (AP selection) but the topic continues to be relevant.
- 'Testing Mobile Data Applications on Smartphoes with SymPA' by Almudena Díaz and Pedro Merino discusses further the shortcomings of operating system platforms for data networking on cell phones and related devices. This is important, since the largest growth in Internet-capable devices is currently in this area.

Chapter 3

Tools and Toys of New Media

Reporters, editors and photographers all store data digitally. Even if it's just your list of contacts, learn to manage your data electronically to maximize its usefulness in the future. And open up your reporting to harness the power of the public. As disruptive as Web sites have been to the traditional publishing and broadcasting model, we ain't seen nothin' yet. The digital landscape is awash in change. Before you learn the basic skills that will allow you to participate in the digital revolution, it's important for you to look at the lay of the land through a broad lens. New and evolving technology and gadgets have changed—and will continue to change—the markets that news operations are aiming to serve.

I will start with information about some tools you should be using and then discuss tools you should understand that others are using. Not everyone wants to watch movies on their cell phones, but there are some very simple tools and practices you should adopt as you become digitally literate.

Tools you should be Using

Vanity searching: If you're in a position to hire others, you have almost certainly added Google and Yahoo! to the screening process. Conducting a Web search of a prospective job candidate is a common step in the early sorting process and the results can be revealing. A 2006 survey conducted by CareerBuilder found that, of hiring managers who used Web searches to research job candidates, 51 percent have eliminated a candidate based on what they found. If they searched a social networking site like MySpace or Facebook, the results were more ominous: 63 percent did not hire the person based on what they found. What does this mean for you? Before you apply for a new job, do some vanity searching in both Google and Yahoo! Make sure there aren't any compromising photos or inappropriate material. And if you're a recent college grad, better check MySpace and Facebook, too. Just because you've never posted a picture of some wild times at a party doesn't mean that no one else has. (If you find something, hopefully you can contact the "friend" who posted it and ask that it be removed.)

Flash drives and memory cards: Remember floppy disks? You probably have a stack of them at home or work, yet you never use them anymore. That's because they hold such a small amount of data—1.4MB—that they're just not practical. Think about it this way: A floppy disk could hold one, maybe two, digital photos. Today's digital landscape relies on small devices with huge amounts of storage. USB flash drives (for text) and memory cards like compact flash (CF) or secure digital (SD) (for digital photos or added game memory) can store hundreds or thousands of megabytes. And, like most technology, the prices on these devices have dropped precipitously since they hit the mainstream. A 1GB flash drive cost as much as $100 in 2004. In 2006, the same drive cost as little as $19. As a result, as many as 150 million flash drives were expected to be sold in 2006.

What does this mean for you? If you work with text, you can feasibly back up all your documents on a flash drive every time you log off your computer. It's quick and easy and, as the saying goes, there are two kinds of computer users: Those who back up their data, and those who will.

Another useful application for flash drives is the transfer of large files. If you have photos or a honkin' PDF that you want to send to someone in the newsroom, give the e-mail server a rest and copy it to a flash drive. The recipient can download it in a few seconds and your IT department will thank you for not clogging the server with several MBs of an attachment. Flash drives have a bright future, too. In September 2006, the USB Flash Drive Alliance announced it will endorse a new generation of "smart" drives that will allow users to run active programs from flash drives. So in addition to document and image storage, the new flash drives will have your Web browser with all your bookmarks, your instant messaging program with all your buddies, your games and more, all encrypted to keep it safe. So wherever you are in the world, you could use any computer and it would be just like using the one at your home or office.

Mobile 2.0

Now that you understand a little more about Web 2.0—and you're reading about Journalism 2.0—it's time to introduce Mobile 2.0. The next generation of wireless connectivity to mobile phones will allow regular cell phones, smart phones, BlackBerries and other devices to connect to the Internet via a high-speed network. Data will transfer as much as 10 times faster, according to some reports, which will make video, music, games and e-mail convenient to anyone, anywhere.

In effect, this is like going from a dial-up Internet connection to a high-speed hookup like the one you have at work.

Even before the third generation (commonly referred to as 3G) saturates the market, mobile delivery is a great opportunity for local publishers and broadcasters. Calendar listings, sports scores, news and weather updates are all within the regular operation of most local news publishing operations. Delivering them to mobile phones and other portable devices is the next logical step. Look at the market: There are 200 million mobile phone users in the U.S. and 70 percent are Web-enabled; 35 percent of those who have the Web option are "regular" users.

- The Weather Channel has 4.8 million paying subscribers a month for its mobile service.
- ABC/Disney has 2 million subscribers at $15 a month delivering ring tones, voice tones, wallpapers and video clips.

You should be aware that dozens of companies are working hard to make what is happening now obsolete, so to preview the promise of any of these technologies would be like predicting the future. Some of these new technologies will gain critical mass and change the world of communications, but, if I knew which ones, I would be a venture capitalist instead of a journalist.

What does this mean for you? The push for immediacy will continue as news operations master breaking news on a Web site and move to present breaking news on mobile devices. It also means a broadening of the scope of information that will be considered worthy of an immediate update, meaning all types of information and news (sports, business, entertainment) will be part of the mobile equation.

iPod: The Slim, Sleek, 800-pound Gorilla

One gadget that has already changed the media landscape is the Apple iPod. By describing the capabilities and uses of the iPod here, I mean to include any of the MP3 players on the market with video capability. No other device has changed the media landscape like Apple's player and iTunes stores.

As of November 2006, Apple had sold nearly 60 million iPods in the five-year life of the gadget, with 36 million sales in the past 12 months. According to Piper Jaffray & Co. research released in October 2006, the iPod owns 79 percent of the market share for digital media players. And talk about a youth market. Other Piper Jaffray research on teenagers found that 72 percent own an MP3 player and 79 percent of those specifically own an iPod. Almost half of the 1,000 students surveyed expect to buy a new media player within a year, and 76 percent of those prefer the iPod.

Some mainstream media companies are responding to this growing market. In September 2006, ABC News began creating a daily 15-minute newscast, separate from "ABC World News," frequently using the same anchor, Charles Gibson. The "World News Webcast" is available through the Web site at 3 p.m. ET and ready to download on iTunes about an hour later. There were more than 5 million downloads in both September and October, 2006. Newspapers such as the Roanoke Times in Virginia and the Naples Daily News in Florida began "vodcasting" in 2006. Each paper has built a studio for recording and producing video segments and each is making those shows available for download to an iPod or viewable on the Web site. National Public Radio, meanwhile, serves more than 6 million downloads of its podcasts each month. What does this mean for you? Every news organization is likely to try adding video to its mix very soon (if it hasn't already). If you can be an early adopter and find a way to incorporate video into your beat or your specialty, you will have a leg up on the competition.

'Other' Wireless

Some people actually still connect to the wireless Internet with a laptop computer. OK, that's being too flip, since laptops are still the primary vehicle for people to use with the Internet, but when you see what's happening with iPods and cell phones, it's easy to forget. Hitting a coffee shop with your laptop and paying a few bucks to connect to the Internet is one of the most popular ways to work wirelessly these days. But that business model doesn't look promising. Independent coffee shops, restaurants, car dealers, rock-climbing gyms and all sorts of other small businesses now offer free Wi-Fi access, too. And the field is only getting more crowded.

- Many cities are working on municipal Wi-Fi systems to bring free wireless Internet access to a concentrated area like a downtown.
- Special cards provided by the major cell phone companies insert into most laptops and allow wireless connection to the Internet from anywhere there's cell phone coverage. Users pay for the

card—usually less than $100—then pay a monthly service fee for unlimited connectivity. A new service, called EV-DO, offers broadband-like speeds.

- A company called Clearwire, founded by Craig McCaw, who built one of the first cell phone companies, is launching in several U.S. cities. It offers standard wireless service or a special modem-like device that can be plugged into a laptop or desktop computer for more reliable service at a higher speed. The idea is that you could pay for one service that would go with you anywhere, so Clearwire would be your provider at home, at the local coffee shop, or wherever. You would use this external modem to connect at home or take it into coffee shop and use it instead of paying six bucks for a T-Mobile hook-up. It's like having your home Internet connection anywhere.

What does this mean for you? The online audience served by breaking-news updates throughout the day will continue to grow. Thought of as the "at-work" audience for much of the digital age, potential readers of news updates will grow as wireless Internet service becomes free and ubiquitous. Combined with Mobile 2.0 gadgets and services and the continued mainstream adoption of downloaded material on iPods, the opportunities for news companies to reach customers digitally will continue to explode.

Get your "mojo" on: The increasing adoption of mobile communication technology not only changes the way audiences receive the news, but also opens up new ways to report it. Mobile journalists—or "mojos"—are becoming more common at TV news stations and even popping up at newspapers. Also known as backpack journalists, these multidimensional dynamos can carry an assortment of tools into the field to report the news in a fully multimedia manner. A laptop with wireless Internet connection, a video camera (that also shoots still photos) and an audio recorder are the basic pieces of equipment that allow journalists to produce news stories or blog posts, photos, video, or audio for a story. Yahoo!'s Kevin Sites is the best-known backpack journalist working today. Sites travelled to war-torn countries around the world to tell the stories of those most affected by calamity, and his regular feature, In The Hot Zone, on Yahoo! News claimed 2 million readers a week in 2006.

The News-Press in Fort Myers, Fla., meanwhile, dispatched several "mojos" into the field every day to report hyperlocal news close to home. These new era reporters have no desks and rarely a specific assignment outside of a geographical area to visit. They drive around their area and perform a modern version of that "old shoe leather reporting." They also do marketing, handing out fliers to educate people about the news organization's online services. Frank Ahrens of The Washington Post profiled the News-Press mojos in December 2006 (and shot this photo of Kevin Myron in his car). "Their guiding principle: A constantly updated stream of intensely local, fresh Web content—regardless of its traditional news value—is key to building online and newspaper readership." Mojos are still an experiment, but if nothing else, they show how flexible—and mobile—journalists in the future can be when it comes to covering and reporting news.

Digital Audio and Podcasting

Eventually you will be asked to capture audio to go with your story (if you haven't been already). Learn the basics of gathering natural sound, recording an interview and editing the clip (with free software) in this chapter. A challenge for many reporters is to capture in words a story's particular sights and sounds. Photographs usually solve the visual end of this equation. Now, with the advent of cheap digital audio recorders, reporters can bring readers even closer to the story by enhancing their reporting with audio clips.

The Basics: Audio Formats

It's helpful to have an understanding of digital file formats as you get started. If you download or listen to audio on a Web site, then it is probably in a compressed format so that it downloads faster. You're probably familiar with some of the formats, like MP3 and Windows Media. It's not necessary for you to know the technical differences between them, just know what you're dealing with. Here's a glance at the most prevalent formats of digital audio.

Compressed (on Web sites);

- MP3 (most universal)
- WMA (Windows Media)
- Real (Real Audio)
- MPEG-4 (QuickTime)
- MPEG-4 AAC (iTunes).

Uncompressed (not found on Web sites)

- WAV (pronounced "wave")
- AIFF (Apple's standard format).

Your goal should be to provide audio clips in MP3 format for your readers. Why? Because virtually any computer can play an MP3. Programs like iTunes, Windows Media Player or Real Player can play them, too, but they can't play the other proprietary formats. For example, you can't play a Windows Media file in iTunes or a Real Media file in Windows Media Player, but you can play an MP3 on any of them.

Identifying Opportunities

If you are a reporter, interviewing people is what you do. Sure, you can transcribe the best quotes for print but does that really provide a thorough and complete report? Did one of your sources elaborate on an important topic, which you then paraphrased to avoid a long quote? Did someone say something with emotion or feeling or uniqueness that doesn't transfer to text? Most news articles can be improved with the addition of audio clips. A newspaper reporter can easily produce audio clips on more than half of the stories he or she turns in, based on the subject matter alone. That may sound too ambitious if you haven't edited and published audio for the Web before. But once you do it a couple of times, it will become second nature. The first step is to toss that microcassette machine from the 1990s and get yourself a digital recorder.

Buying A Recorder

Like most digital tools, there are plenty of options on the market today and deciding which one is right for you starts with a basic question: How much can you afford to spend? You can buy a new digital recorder for as little as $50, but if you spend even a little bit more you can dramatically improve your capacity to participate in this game. And, of course, if you spend even more you will go from "entry level" to "professional" in no time.

The key points to consider include recording time, digital file format and compatibility with your computer, ease of use and transferring files, and quality of recording. Let's look at a few options in different price levels and evaluate them on these aspects. Important note: You might be tempted to buy a $50 version because, hey, it says it's a digital recorder. But unless you can transfer the files from the recorder to your computer, you will be unable to get the files onto a Web site where readers can listen to them. So it would be like writing a story on a computer and not being able to send it to your editor.

Using A Microphone

While using an external mike can be an extra nuisance during an interview, the added sound quality is well worth the effort. There are basically two types of external microphones: A standard mike with a cord, and a wireless or lavalier mike. Let's take a look at the advantage both have to offer and also explore the best way to record telephone calls digitally.

A standard mike with a cord is helpful if you are interviewing more than one person at a time or you want to include your voice on the audio clip so listeners can hear the full interview instead of just selected quotes. It is also the best way to gather natural, or environmental, sound, which can be spliced into the audio segment to enhance the listening experience. Gathering natural sound is not the same as background noise. Interviews should be done in a setting that allows the voices to be recorded without interruption. Separate from the interview session, however, it's always a good idea to search for those sounds that will help describe the setting. Are there power tools being used? Is it a noisy office with lots of chatter and phones ringing? Is it an outside setting where you can hear the bugs and the birds?

If there is natural sound to be had, take just a few minutes and record it—without anyone talking. "You might feel silly just standing there holding your mike in the air, but when you get back to edit your stuff, you'll be glad you have it," said Kirsten Kendrick, a reporter and morning host on KPLU radio, an NPR affiliate in Seattle and Tacoma. You should record natural sound in uninterrupted 15-second increments. That way you'll avoid the problem of not having enough to use in editing. You can always make a clip shorter by cutting it but you can't make it longer, so make sure the material you're working from is long enough to cut from.

Assignment

Find the NPR station near you or listen online from NPR's Web site. The public radio broadcasters do a masterful job of weaving natural sound into their reports. And as a listener, you get a better sense of the setting for the story when you hear what it really sounds like. A wireless or lavalier mike is most helpful when your goal is to capture the voice and words of one person and you're "in the field." While they might be intimidating at first, wireless mikes are really very simple. There are two halves: A battery pack and miniature mike on a cord that clips on the person you want to record (this sends the signal), and a battery pack and cord that goes into your recording device (this receives the signal).

Here's how to get started:

1. Clip the mike on the lapel of your subject and give them the battery pack to put in their pocket. Don't forget to turn the device on!

2. Connect the receiver pack to your recording device, turn it on, and put it in your pocket or purse or handbag. Then operate your recorder as you normally would: Hit the record button when you're ready and pause button if there's a break in the action.

Recording with Your Computer

To record a phone call digitally, you'll need another piece of equipment: A telephone recording control unit that sells at Radio Shack for $25. Many journalists already use one of these to record phone calls to their analog microcassette tape recorder. And those same journalists probably have an unruly jungle of tapes on their desk or in a drawer that is unlikely to produce the tape from six months ago that someone might need. (NOTE: In some states it's illegal to record someone on the phone without their express permission.)

That's one reason to go digital—organization. With the Radio Shack device you can record directly to your computer, which makes it easy to store files in an organized fashion. And going through the "tape" is easier on a computer since most playback programs like Windows Media Player have slider bars that allow you to quickly go from the beginning of a recording to the end.

Your hardware is ready. Now you need software to manage and edit the sound files with your computer. There are literally hundreds of options for audio software, ranging from Adobe Audition (the choice for most radio professionals, $349) to Audacity and JetAudio, popular free downloads that work great.

No matter which audio software you use, there are a few basic settings on your computer to check before starting your first recording:

- *File name:* You will either need to select File-> New and create a file or choose where on your system this new file will be created. Either way you need to think about what to call your file. This is a good time to come up with a standard file naming convention that will serve you for months and years to come. Include the date and the name of the person you'll be talking to, so an interview on Valentine's Day with Paris Hilton would be named "021407hilton." It's also helpful to create new folders by year or month for more organization.
- *Format:* You should record in WAV format so your files are uncompressed and, therefore, of the highest quality. You can convert the files to MP3 (Audacity and JetAudio can both do

this) once they're edited for publishing on the Web. You only need to worry about this when recording directly into your computer, not when using a digital recorder.

- *Input/Mike level:* Make sure the software is set to capture data via microphone input. Then find the setting that adjusts the microphone level and set it to about 70 percent of the possible level.

Assignment: Now call a friend and record the call for a trial run. Save the file with your new naming convention. Play it back to make sure it sounds good.

Editing Your Audio

It's unlikely you'll ever publish an entire session online. Just like you don't publish entire interviews in text, you need to edit your audio to make sure the best stuff is not obscured by less compelling, less important or repetitive content. Editing audio is remarkably similar to editing text, so you shouldn't be intimidated when approaching this task. First, acquire the audio file if it's still on your recording device. Connect the digital recorder to your computer through USB and drag the file(s) you need into a folder or onto the desktop. Important note: Most—but not all—digital recorders come with a USB cord that connects to a computer to make this easy (simply plug the cord into the recorder, then connect it to the computer through a USB port). The cheapest recorders, however, do not interface with computers, making them much less useful.

Launch your audio-editing software. Ideally, the program should be easy to use and export files in MP3 format. If you use a PC or a Windows machine, Audacity and JetAudio are excellent free options. Let's go through the editing process with Audacity since it appears to be the most prevalent free software in use today.

Editing with Audacity

1. Use File-> Open and open the audio file
2. *Crop out the bad stuff:* Think about how users would best appreciate the content—in one full serving or broken up into smaller bites. Highlight areas that represent unwanted ums, ahs, mouth noises and lip smacking. Then simply hit DELETE. Also crop out silence and any small talk at the beginning and end.
3. *Make it stereo:* Some files will be mono, not stereo, meaning you'll only hear the audio in one side of your headphones. You want to make it stereo so the sound file will play in both sides of speakers and headphones, instead of just one. To make it

stereo, click on the Audio Track label next to an upside down triangle. Then select Split Stereo Track from the drop-down menu. Then copy the region that you've edited by highlighting it and using Edit-> Copy. Then click into the lower window and use Edit-> Paste.

4. *Export the file:* Convert your audio edit into a compressed, ready-for-online publishing MP3. Just go to File and select Export as MP3. Ignore the meta data interface (Author, Description, etc.) unless you're doing a podcast.

Assignment: Record your own voice as a test. Count from 1-10 into a microphone and capture it digitally. Then edit your take. Highlight the section where you say "3" and select Edit-> Cut. Then move the cursor to after the "6" and select Edit-> Paste. Repeat a few more times with other numbers. This will give you a feel for how the sound waves represent words and sounds and also show you how easy it is to edit audio.

Using Time Points for Speed

Most newspaper journalists will do what they know first—use the audio to get quotes so they can write their story before they edit the audio for online publishing. That's great. But think about the audio editing you will do next as you listen to the entire take. If you make a note of the time when a good quote plays, you'll save loads of time when you go back to edit the take for the good stuff.

All audio-editing software programs feature convenient time track marks, so if your interview's best quote occurred 10 minutes into the interview, you write "10:00" next to the quote in your notebook. Then go directly to the 10-minute mark on the track when you're ready to edit and you've saved yourself 9 minutes, 59 seconds.

Ready for Podcasting

Podcasting is the distribution of audio files over the Internet using RSS subscription. The files can be downloaded to mobile devices such as MP3 players or played on personal computers. The term podcast, (Playable On Demand + broadcast) can mean both the content and the method of delivery. Podcasters' Web sites also may offer direct download of their files, but the subscription feed of automatically delivered new content is what distinguishes a podcast from a simple download. Usually, the podcast features one type of "show" with new episodes available either sporadically or at planned intervals such as daily or weekly.

Podcasting with video files is often referred to as vodcasting (video + podcasting). It works the same, but includes video. If you download a vodcast on an MP3 player that doesn't have a video screen, you will still be able to hear the audio. In format, podcasts are similar to conventional radio programming with a host or hosts interviewing a subject, playing music or introducing pre-recorded audio stories. So it's no surprise that National Public Radio produces some of the most popular podcasts online.

The San Francisco Chronicle was producing two dozen podcasts as of January 2007 on topics as diverse as the San Francisco 49ers football team, wine and movies. Listening to podcasts in iTunes: If you have iTunes, finding and listening to podcasts is simple. Just click the Podcasts link in the left menu, then click Podcasts directory on the bottom of the screen. Search by category or most popular. Click Subscribe if you'd like to add a podcast to your collection and it will automatically update any time there is new content.

Setting up a podcast: If you know you will have regular audio files on a specific topic to offer to readers, setting up a podcast will make organizing and publishing the audio convenient for you and your readers. A good example is a sports beat writer who records interviews with coaches and players and wants to offer them to readers. Setting up a podcast will allow a reader to subscribe and automatically receive new files as they become available. Creating a podcast that others can subscribe to is easy and free—if you have an RSS feed set up. Go to iTunes and click on the Submit a podcast logo or use another service like Podcast Alley.

Functions of Communication

Mass media has a prominent role to play in modern society. It can bring about radical changes and improve social situation as it influences our social, civil, cultural, political, economic and aesthetic outlook. Modernization has converted media into an indispensable feature of human activity. However, factors like age, education, economic condition, personal needs and availability of proper components decide the quantum and frequency of media use. This is evident from the fact that most media centres are located in urban areas. The majority of consumers of media products are also concentrated in and around cities and towns. It is rightly said that media use is an index of development. The greater the use, the higher will be the level of education. As social beings, humans are sustained by mutual interactions, exchange of ideas, information and views with the fellow beings. Illiteracy, which is nothing but absence of education and information is a stumbling block for any aspect of development-social, economic, political, cultural and even spiritual. Media has become the harbinger of development through the removal of these roadblocks and the provision of information and knowledge. In a democratic country like India, the ultimate power lies with the people. But a democratic society needs vigilant and informed people who are able to see through the gimmicks of political parties and politicians. Media creates such valuable citizens.Besides, media has done much good to society by exposing various scams, scandals, frauds, embezzlements and many other cases of corruption leading to initiation of enquiries and other processes of prosecution against the perpetrators of these crimes.

History is witness that press has been instrumental in putting an end to atrocities and bringing the downfall of ruthless dictators. In India, vernacular press did the job of uniting people against the oppressive British rule and triggered its end in the country. However, media too suffers from some pitfalls, growing consumerism and materialism have adversely impacted our media. The partisan attitude, sectarian outlook and biased individualism in some sections of media are a testimony that media too is susceptible to harmful influences. Often, in fierce rivalries, ethics of journalism are thrown out of the window to settle old scores. Running after opportunistic gains is another malady our media suffers from. The incidents of throwing are against the ethics of media. Deliberately creating sensational stuff to attract with reality-is another tactic that media must avoid.

It is said that big minds discuss issues but small or swallow minds discuss/attack persons. The significance of communication for human life cannot be overestimated. This is true because beyond the physical requirements of food and shelter man needs to communicate with his/her fellow human beings. This urge for communication is a primal one and in our contemporary civilization a necessity for survival. That is to say without communication no society can exist, much less develop and survive. For the existence as well as the organization of every society communication is a fundamental and vital process.

Mass Communication and the Mass Media

Among the four identified forms of communication is mass communication, which deals with messages addressed from one to many persons mediated by elements in mass media such as radio, television, film, newspaper, magazine, book etc. Fortunately and not boastfully I can say that since the early 1970s when I became a trained teacher, I have been involved with all the elements of mass media except film. In the 1970s I was writing for radio and television (news commentaries, reflections, and news reports as well as poems and youth programs) in Ghana and the most popular newspaper in Ghana-the Daily Graphic.

By 1980 I had become a regular contributor to the Network Africa program on the B.B.C. World Service. My love for the mass media sent me to the Kwame Nkrumah University of Science and Technology (KNUST) in Kumasi, Ghana to do a degree program in Publishing Management (Book Industry) from 1989 to 1992. I specialized in editing and marketing and worked as an editor for the newsletter of the Department of Book Industry as well as an assistant editor for the students? magazine published by KNUST. Above all, I single-handedly initiated and published a church quarterly paper-the Southern Trumpet-for the S.D.A. Church in South Ghana for almost ten years before leaving Ghana in 1995 for further studies at the University of Toronto in Canada. Since 1998 that I joined the editorial board of the dynamic community newspaper-the Ghanaian News-published monthly in Toronto I have served in the capacities of assistant editor and contributing editor till now. I have therefore come a long way living through the bad and good days of the mass communicator associated with the print media, especially newspapers. The joy of disseminating good news, the anxiety of hitting scoops in news reporting, the embarrassment of the printer?s devil, which causes elementary errors in published articles and the wild attacks on opinions expressed in the mass media! My contribution to the discourse on the role of the mass media in community development is authentic since I understand the nitty-gritty of being a mass communicator.

Mass communication is the technological means of sending information, ideas and opinions from a mass communicator to a complex audience. It is also defined as comprising the institutions and techniques by which specialized groups such as broadcasters, film producers and publishers employ technological devices to disseminate symbolic content to large heterogeneous and widely disperse audience. Mass communicators are impersonal. They are part of the institutions they work for and should not be blamed personally for what comes from the institutions. The credibility of the message is not for the individual communicator, but for the institution or the organization that sends it. Thus, mass communication deals with collective sender. For example, a newspaper is not produced by only one person. The newspaper is the end results of collective efforts of reporters, editors, type-setters, proofreaders, designers and printers. These must be well-trained persons. It is, however, sad to admit that the mass media have been infiltrated by some unqualified persons as well as some unscrupulous individuals with their own political agendas other than to serve the communities they are supposed to educate, inform, entertain and mobilize for development. Hence, the establishment of media commissions in most democratic and civilized communities! Institutions and organizations engaged in mass communication anywhere must be weary of such persons.

The other important area of notice is that mass communication deals with the concept of mass audience. Here, there is no common motivation. Audience do not act together. They belong to different classes-different education and socio-economic status. The message communicated in the mass media is open to the public and everyone has access to it provided she/he has the mass communication technological device as well as understands the language in which the message is sent.

Chapter 4

Computer-Aided Teaching and Learning

The State of Education and Reform

To many observers, the report A Nation At Risk: The Imperative For Educational Reform was a watershed moment in the U.S. debate on education reform. Prior to 1983, reform movements and debate about the performance of the education system were a steady drum beat in the U.S. socio-political consciousness with peaks of concern and activity after WWII; the Civil Rights Movement; and the Soviet launch of Sputnik.

The report was commissioned by then-Secretary of Education T. H. Bell to address "the widespread public perception that something is seriously remiss in our educational system." Blunt and damning in its language, the report states, "If an unfriendly foreign power had attempted to impose on America the mediocre educational performance that exists today, we might well have viewed it as an act of war."

A sampling of specific observations includes:

- "International comparisons of student achievement, completed a decade ago, reveal that on 19 academic tests American students were never first or second and, in comparison with other industrialized nations, were last seven times.
- About 13 per cent of all 17-year-olds in the United States can be considered functionally illiterate. Functional illiteracy among minority youth may run as high as 40 per cent.
- There was a steady decline in science achievement scores of U.S. 17-year-olds as measured by national assessments of science in 1969, 1973, and 1977.

- Between 1975 and 1980, remedial mathematics courses in public 4-year colleges increased by 72 per cent and now constitute one-quarter of all mathematics courses taught in those institutions."

In the wake of A Nation at Risk, we have witnessed a flurry of reform movements, a profound increase in political activity on education reform at the national level in addition to state and local efforts, and an intensification of debate at all levels. What follows is a few of the many education reform activities since 1983.

School Choice

What many refer to as the "School Choice Movement" has gained momentum since the late 1980s. School voucher programmes, first proposed by Milton Friedman in 1955, began in earnest at the local level in 1990 in Milwaukee and the federal level in 2004 with the creation of a programme in Washington, D.C. The first charter schools were founded in Minnesota in 1991 and now educate more than one million students in 3,600 schools in 40 states. While firm statistics are difficult to obtain, an estimated 1 million students in the 2005–2006 school year were educated in home schools up from approximately 15,000 in the early 1980s.

National

Government and legal activities in education reform have occurred on a wide variety of fronts. The "Even Start" programme was approved by the U.S. Congress in 1988. This programme combined both improved early childhood education with adult family literacy in an attempt to end the generational cycle of illiteracy in the U.S. In 1989, the Kentucky Supreme Court declared the entire state education system unconstitutional on the basis that the state was not providing minimally adequate education to which public school students were constitutionally guaranteed. As of 2000, twelve other states including New Hampshire, Wyoming, Alabama and New Jersey faced and lost similar "adequacy-based" lawsuits. In 1994, the U.S. Congress enacted "Goals 2000: Educate America Act", a bill intended to improve teaching and learning by creating a national framework for education reform, encourage high levels of academic achievement and "promote the development and adoption of a voluntary national system of skill standards and certifications".

Also in 1994, Proposition 187, passed in California in a general election referendum, banned undocumented immigrants from public education. In 1996, President Clinton in his State of the Union address, called for an end to the practice of "social promotion", a practice in which students are allowed to progress to the next grade level regardless of actual academic achievement. And in 2001, the U.S. Congress passed the "No Child Left Behind" Act which, among other mandates, transformed voluntary "Goals 2000" national education standards to compulsory ones and made federal funding conditional on school performance. This "accountability" feature in NCLB reflected a growing trend through the 1990s at the states-level of instituting or expanding accountability systems.

State and Local

Across the country, numerous small-scale programmes, teaching movements and pilot programmes have been altering the education landscape. Overall, state-level reform efforts can be characterized as focusing on education outcomes. As a result, nearly all states have raised their academic standards and instituted new assessment programmes. As of 2004, twenty states added exit examinations to their high school graduation requirements up from 10 in 2001. A perennial plank in the National Educators Association's platform has been reducing class size at all grade levels. Beginning in 1985, Tennessee's Department of Education launched the four-year, Student/ Teacher Achievement Ratio programme to study the effects of classroom size on student learning at the kindergarten through 3rd grade.

A follow-up study was conducted on the cohort in the 10th-grade. Wisconsin began the Student Achievement Guarantee in Education programme in the 1996- 1997 school year, a programme aimed at maintaining student-teacher ratios at 15-1 or below for kindergarten through 3rd grade. This selection of reform efforts is by no means comprehensive. Reform developments in pedagogy, technology, teacher education and qualification, school and classroom architecture, special education, higher education and more are clearly missing. Even so, this brief presentation demonstrates the intensity of education reform since 1983. After so much work, where are we now?

National Education Trends

The National Center for Education Statistics, in conjunction with the Institute of Education Sciences and within the auspices of the Department of Education, is the primary entity for collecting and analyzing education data for the federal government. NCES maintains two parallel programmes for measuring national progress in education: the National Assessment for Educational Progress Long-Term Trends programme and the "main" NAEP assessments programme. The LTT programme has employed the same assessment instruments and methodologies since inception, allowing for the direct comparison of results from year-to-year. The main NAEP programme is subject to revisions as new priorities, content and methodologies emerge. Because the two programmes differ significantly, it is not possible to directly compare results between programmes.

The points along the trend lines are national averages of scale scores for ages roughly corresponding to 4th, 8th and 12th grade. You may have noted that the trends express little change between the start and end points and minimal variation across the periods with three exceptions: mathematics scores for ages nine and thirteen show twenty-two and fifteen point gains respectively from 1973 to 2004; reading scores improved by nine points from 1971 to 2004. Mathematics, science and reading represent "core" subjects in the NCES assessment strategy. However, NCES does conduct periodic assessments of other subjects including U.S. history, geography, art and music.

The Condition of Education Report 2003 and contain results from the first studies conducted in 1994 in U.S. history and geography and the most recent results from 2001. The methodology of these assessments is closer to the main NAEP than the LTT. The full bar in each column represent 100 per cent of each sample. Like the LTT graphs, you may have observed modest changes between the two sets of assessments. That in all grades, approximately one-third of all students were below basic standards in these subject areas with the exception of 12th grade U.S. history achievement in which over one-half were below basic standards.

As discussed earlier, the main NAEP assessments evolve over time. Because the assessment methodologies remained relatively consistent from 1990 to present in mathematics and from 1996 to present in science, year-to-year comparisons can be made with a significant degree of reliability and validity. The primary features are achievement levels by grade expressed as percentages of each sample. However, these express achievement levels as groupings, i.e. at or above Basic, etc., a style of presentation seen in most of the recent NAEP literature. Mathematics progress for the 4th grade shows significant improvement with a twenty per cent decrease in below Basic achievement from 1990 to 2005.

The data also suggests acceleration in the improvements made at the 4th grade level from 2000 to 2005. Smaller but noteworthy gains are exhibited in the 8th and 12th grade samples. Note the correspondence in trends between the main NAEP mathematics data and the LTT graphs. However, take notice of the fact that below Basic achievement in 2005 still ranges from twenty per cent to thirty-five per cent. In contrast, science achievement reveals little or no improvement from 1996 to 2005 and below Basic achievement ranges remain in the thirty to forty-five per cent range. There is also correspondence between the main NAEP Science and the LTT Science trends although a comparison of three assessment intervals to two assessment intervals is far from conclusive. An interesting pattern, or perhaps coincidence, emerges from a comparison of all three sets of data: 12th grade/age 17 academic improvements are the least significant between the three groups. Is this pattern real or a statistical anomaly? If the pattern is real, what might be the cause? A quick review of data from Kentucky and Colorado suggests that this pattern might be born out at the state level, too. This observation merits further study. Switching gears, let us consider the premise that there was a net increase in the quantity of education reform since the publication of the A Nation at Risk report in 1983. To be clear, TCFIR is not yet aware of any study supporting this premise empirically.

Rather, this premise is based on an impression formed in our own research of the literature. Let us also acknowledge that the phrase "education reform" suffers from a lack of a specific and generally agreed upon definition which further weakens the premise. However, it can be said with certainty that the statement "No education reform has taken place since 1983" is false. Let us consider a second premise: The education reforms that have been put into service since 1983 have been executed in good faith, backed by sound research and an expectation of success.

Finally, let us consider the third premise: The assessment instruments and research programmes evaluating both educational achievement and the effects of education reform have external and internal validity. The third premise is supported by at least one study. Let us consider an addition premise: There is no one "right" reform of education, no "magic bullet", nor can all reforms be "New Math". The effectiveness of reforms should be expected to fall along a range from relatively ineffective to relatively effective regardless of the good faith intent of the designers of reforms. In fact, the effectiveness of the population of reforms would probably fall within a normal distribution if a metric for effectiveness could be created.

But, given premise two, we might presume a negatively skewed distribution so that the mean effect of all reform efforts nationwide would result in an increase in the mean academic achievement. Simply stated, assuming that most reforms are at least mildly effective we should see some general improvement in academic achievement. Given these premises–and even allowing that the pace of education reform was unaffected by A Nation At Risk–do the graphs and reflect the kind of academic achievement gains we would expect to see based on the sheer scale of efforts nationwide to reform education? Are we seeing an adequate return on our education investment? And if we are not, then why not?

Kentucky and the KERA Reform Effort

Can we see evidence of a positive effect of education reform on achievement at a smaller scale? In 1989, the Kentucky Supreme Court ruled in the case of Rose v. Council for Better Education that the state's education system was unconstitutional. The suit contended that the education system failed the test of "efficiency" and "minimally adequate education" for all of the state's students. The Court ordered the state bring the education system into constitutional compliance within one year, forcing one of the most sweeping overhauls of a state education system in U.S. history. The Supreme Court took the unusual step of enumerating the characteristics of an efficient school system in eight of the pages within the decision.

The Kentucky Education Reform Act of 1990 "recreated the entire education system and included not only finance and governance changes, but also programme changes". KERA established new education standards and the Kentucky Instructional Results Information System assessment. The statewide education budget was increased by 32% from 1990 to 1999.

Other effects and mandates of KERA included:

- Creation of the first online university in the nation.
- Decentralized decision making with major decisions shifting to the school level. The objective was to improve parent involvement and parent-teacher cooperation in issues related to instruction, curriculum and even personnel.
- The creation of a primary school programme to replace the kindergarten through 3rd grade system that included multi-age, multi-ability classrooms.
- Major changes to teacher training and qualification requirements including the creation of a mentoring programme for all novice teachers.

KERA has drawn both strong praise and strong criticism. In the mid-1990's, the Ford Foundation and Harvard University awarded Kentucky the Innovations in Government Award, Kentucky will receive

2006 Frank Newman Award for State Innovation from Education Commission of the States. The Department of Education has cited Kentucky in several NAEP reports for significant improvements in its mathematics and science education scores.

Inspired by the precedent set by the Rose v. Council for Better Education decision, twelve other states have faced similar lawsuits and a number of states and municipalities including Arkansas and New York City have looked to Kentucky as a potential model of education reform. On the other hand, critics have sited continuing inequities in funding due to formulas that fail to adequately fund schools with relatively small reduced- or free-lunch student populations. A study in 2004 "suggested that Kentucky schools need $740 million to $1.2 billion more a year, on top of their current $4.1 billion allocation, to adequately fund the promises of KERA" while parties from the Rose v. Council for Better Education suit have threatening new legal action based on these same funding concerns.

Some education reform observers suggest that Kentucky now leads the nation in the rate of increase in non-public school enrollments, indicating that parents are giving up on KERA reforms. Another resource suggests that the Kentucky Department of Education was warned by American College Test, Inc. of problems with the reliability of KIRIS but suppressed the information until pressed by state lawmakers. Other questions have been raised about the reliability and validity of KIRIS especially in comparison to results from NAEP. In both cases, the implication is that KIRIS is overstating education achievement in the state. There are criticisms about Kentucky HB 178, a bill which redefined students awarded GEDs as successful graduates of the education system. Observers claim that the bill is in conflict with the stated goals of both KERA and NCLB and represents a manipulation of the statistics on the Kentucky Department of Education's actual performance. One report states, "Lawmakers are so concerned about the department's reports that they ordered an audit of Kentucky's dropout data in August 2003." While NAEP data indicates a clear trend of improvement for Kentucky, the results can be characterized as bringing Kentucky within the range of the national scale score averages. The rate of gains over fifteen years in academic achievement place Kentucky in the top one-third of states in mathematics and science but in the middle of the pack in reading. Perhaps more importantly, recent data shows a significant slowing in academic gains. As a "model" programme, the Kentucky example raises important questions about the effectiveness of current education reform strategies.

International Comparisons

We examined the effects of U.S. reforms represented in the national data and, on a smaller scale, the results of one of the most extensive state-level reforms in U.S. history: the KERA programme in Kentucky. Let us turn our attention to international comparisons of academic achievement. In the 20th century, the U.S. achieved an unparalleled global economic and political dominance due, in part, to high academic achievement relative to other nations and leadership in science, engineering and technology.

With the advent of globalization and profound investments on the part of other nations such as Ireland, India, China, Singapore, Taiwan and Japan in the improvement of their education systems, the relative advantages in education that the U.S. once enjoyed have narrowed. Concerns in the U.S. over relative education quality are not new as the Sputnik-era reforms demonstrate. However, the problem of relative U.S. educational achievement has become especially acute as both the Indian and Chinese economies, representing over 2.5 billion people combined, grew in 2005 at a rate 7.6% and 9.9% respectively and both countries begin to supplant U.S. dominance in technology and manufacturing. In light of these global changes, the Department of Education has participated in a number of programmes to assess the relative quality of the U.S. education system. One such programme, the Trends in International Mathematics and Science Study is conducted by the International Association for the Evaluation of Educational Achievement. The third and most recent study was conducted in 2003 and is summarized below:

Mathematics:

- At the 4th grade level, the U.S. was 12th out of 25 countries with a score of 518 compared to the average score of 495 and the high score of 594 from Singapore. The U.S. score was unchanged from 1995. This, however, represents a reduction in relative standing.
- At the 8th grade level, the U.S. was 15th out of 45 countries with a score of 504 compared to the average score of 466 and the high score of 595 from Singapore. The U.S. score improved by 12 points from 1995 and represented an improvement in relative standing.

Science:

- At the 4th grade level, the U.S. was 6th out of 25 countries with a score of 536 compared to the average score of 489 and the high score of 565 from Singapore. The U.S. score dropped 6 points from 1995 and represented a lower relative standing.

- At the 8th grade level, the U.S. was 9th out of 45 countries with a score of 527 compared to the average score of 473 and the high score of 578 from Singapore. The U.S. score improved by 15 points from 1995 and represented an increase in relative standing.

Reading literacy has been studied in a separate effort. The Progress in International Reading Literacy Study was conducted by IEA in 2001 as the first in a five-year cycle of trend studies on reading literacy. The last comparable study was conducted in 1991 by IEA. PIRLS was administered in 35 countries to 9 year olds. PIRLS is designed to measure reading ability related to enjoyment and reading to acquire and use information, specifically the ways in readers construct meaning from text. The scale was set for 1000 with an average score of 500 and a standard deviation of 100.

A summary of results are as follows:

- The U.S. ranked 9th in combined literacy with a score of 542 compared to the high score of 561 by Sweden.
- The U.S. ranked 4th in the literary subscale with a score of 550 compared to the high score of 559 by Sweden.
- The U.S. ranked 13th in the information subscale with a score of 533 compared to the high score of 559 by Sweden.
- U.S. Black and Hispanic combined literacy scores of 502 and 517 respectively were significantly below the U.S. average of 542.
- U.S. schools with a student body consisting of 75% or more free- or reduced-price lunch recipients had a combined literacy score of 485 compared to the U.S. average score of 542.

As with the national data from NCES, the IEA assessments present a mixed picture of educational reform progress in the U.S. Most importantly, especially in light of the imperative of maintaining a strong position for the U.S. in the global economy, the U.S. does not lead in any of the three indicators of relative academic achievement. The literacy data also highlights the problems the U.S. has in addressing gaps in student education across ethnic groups. As one of the most ethnically diverse countries in the world, the failure to achieve parity in education among all ethnic groups represents at best a waste of "human capital" and a pivotal problem that the U.S. must overcome.

This year, the National Academies published Rising Above the Gathering Storm: Energizing and Employing America for a Brighter Economic Future, a report commissioned by the Energy Subcommittee of the Senate Energy and Natural Resources Committee in May, 2005 for the purpose of assessing the current state of U.S. science and technology and making federal policy recommendations. The U.S. education system was one of several domains investigated by the National Academies.

Among the report's findings were:

- "In 2003 the Organisation for Economic Co-operation and Development's Programme for International Student Assessment measured the performance of 15-year-olds in 49 industrialized countries. It found that US students scored in the middle or in the bottom half of the group in three important ways: our students placed 16th in reading, 19th in science literacy, and 24th in mathematics. In 1996, US 12th graders performed below the international average of 21 countries on a test of general knowledge in mathematics and science."
- "Fewer than one-third of U.S. 4th grade and 8th grade students performed at or above a level called "proficient" in mathematics; "proficiency" was considered the ability to exhibit competence with challenging subject matter. Alarmingly, about one-third of the 4th graders and one-fifth of the 8th graders lacked the competence to perform even basic mathematical computations."

The PISA results are consistent with the IEA studies, consistent with the main and Longterm Trend NAEP assessments and consistent as well with individual state-level achievement data. We believe that it is highly significant that international, national and state-level assessments –using different assessment methodologies–converge on roughly the same appraisal of U.S. academic achievement. The National Academies report provides a body of interesting facts on higher education that depict another dimension of the state of the U.S. education system.

These facts need to be quoted in whole:

- "In South Korea, 38% of all undergraduates receive their degrees in natural science or engineering. In France, the figure is 47%, in China, 50%, and in Singapore 67%. In the United States, the corresponding figure is 15%.
- Some 34% per cent of doctoral degrees in natural sciences and 56% of engineering PhDs in the United States are awarded to foreign-born students.
- In the U.S. science and technology workforce in 2000, 38% of PhDs were foreign-born.
- Estimates of the number of engineers, computer scientists, and information technology students who obtain 2- 3-, or 4-year degrees vary. One estimate is that in 2004, China graduated

about 350,000 engineers, computer scientists, and information technologists with 4-year degrees, while the United States graduated about 140,000. China also graduated about 290,000 with 3-year degrees in these same fields, while the US graduated about 85,000 with 2- or 3-year degrees. Over the past 3 years alone, both China and India have doubled their production of 3- and 4-year degrees in these fields, while the United States production of engineers is stagnant and the rate of production of computer scientists and information technologists doubled.

- About one-third of US students intending to major in engineering switch majors before graduating.
- There were almost twice as many US physics bachelor's degrees awarded as in 1956, the last graduating class before Sputnik than in 2004.
- More S&P 500 CEOs obtained their undergraduate degrees in engineering than in any other field."

The output of the U.S. primary and secondary education system becomes the input to higher education and the workforce. As such, other factors not easily measured by national or state assessments in particular subject areas emerge as trends in higher education and the workplace. One area requiring additional research is how well academic standards and practices actually prepare students for the workplace and post-secondary education. Karl Zinsmeister states, "Employers, too, are distraught. Noting that 44 per cent of the job-seekers who showed up at his office couldn't read at the ninth-grade level, Prudential Insurance executive Robert Winters mourned that 'they are 17 years old and virtually unemployable for life.'" Zinsmeister, 80 per cent of job applicants fail a 5th grade mathematics and 7th grade English competency exam used by Motorola for factory employment.

At the post-secondary level, 72.5 per cent of four-year institutions and 99.4 per cent of "at least two but less than four year" institutions provide remedial education services. In addition, recent reports on high school graduation rates suggest that state and national data is flawed and overstated. And, given the facts provided by Rising Above the Gathering Storm, one must ask what is happening in primary and secondary education to divert graduates from scientific and technical higher education and career choices?

The NCES The Condition of Education 2006 reports that:

- "The number of bachelor's degrees awarded increased by 33 per cent between 1989-90 and 2003-04, while the number of associate's degrees increased by 46 per cent.

- The sole decline among the top five most popular degree fields between 1989-90 and 2003-04 was in engineering and engineering technologies."

What is the number one major in the U.S.? The Job Outlook 2005 survey, the answer is accounting. The issue, then, is not whether U.S. student are seeking higher education. The "leakage" in the pipeline through higher education away from technical disciples has profound implications for the future of the U.S. economy as do the questions about high school graduation rates and basic academic skill levels at the other end of the workplace spectrum.

The National Academies report states that an estimated one-half of U.S. economic growth since World War II has been the result of technological innovation. Based on the rate of natural science and engineering degrees earned in other countries compared to the U.S., the probability of innovation-driven economic growth being centered outside the U.S. is high. The relative strength of the U.S. education system raises troubling questions about the future.

Notes on the State of Education and Reform

For more than two decades, reform of the U.S. education has been a "hot topic" and the focus of great effort at all levels and yet the results as measured by state, national and international assessments are mediocre. NCLB promises to raise academic achievement to nearuniversal standards of "proficiency" but the long-term trends suggest that the goal may be beyond reach within the current paradigms of education even while the short-term trends suggest that improvements are occurring. The sidebar "Then and Now" presents two quotes, one from A Nation At Risk and the other from Rising Above a Gathering Storm; the similarities are disturbing given their separation by more than two decades.

Is it unfair to say that, like Alice in the Red Queen's court, we are running as fast as we can just to stand still? For those who might ask whether we have a problem with our education system, the answer is an unequivocal, "Yes!" The "School Choice" movement is just one example that demonstrates frustration on the part of parents with public education by creating alternatives to a system some say is broken beyond repair, while lawsuits such as Rose v. Council for Better Education attack the system from within. In the realm of the social sciences, public perception is often recognized as a force which creates "facts" regardless of the data. The data on the education system says "maybe" but strong, vocal and active social factions are actively working to reform education outside or on the margins of the system.

This alone creates an imperative to reform education. But the "averageness" of the U.S. education system in international comparisons draws us away from the domain of perception back to objective fact. The U. S. economy has benefited from the disparities between our educational, scientific and technological accomplishments and those of other nations through the 20th century. Other nations, looking to our example, are fast on our heels. If the future trends of the U.S. education system resemble the trends from the LTT data projected forward by thirty years, the U.S. will most surely slip behind. Maintenance of the status quo in academic performance equates to failure on the international stage. From anthropology, we learn that inter-cultural contact, and the resulting diffusion of ideas, technology, etc., is a major driver of cultural change. In this way, U.S. successes have diffused to the world and now we must compete with our own ideas uniquely transformed and reinvented by dozens of countries. The situation is not all bad. Strides have been made in the educations system. Recognition of the importance of early childhood education and programmes such as Head Start has improved the readiness of many children to be educated.

The focus placed on removing institutional and social inequities in the education system have reduced performance gaps between genders and ethnics groups, although those gaps remain. For most of U.S. history, primary control of the education system existed at the state and local level, resulting in a diversity of standards, curricula and infrastructures that created challenges in addressing even the most basic issues of education reform. While there are many criticisms of "standards-based education" and greater federal control of education, the U.S. for the first time has the potential to see the effect of education reforms by reducing this diversity to something more manageable while elevating accountability and measurement of academic performance to new levels of importance. For better or worse, NCLB has put the U.S. on the path of a grand experiment in education. The scale of this experiment will undoubtedly leave its mark in the record of education trends. The need for education reform will not end even if NCLB is a stunning success.

Many observers assert that the world is moving towards a "knowledge-based economy" although that term is as ill-defined as "education reform". What is clear is that technology is transforming every aspect of the world and will continue to do so. Human knowledge continues to grow at a geometric, if not exponential, pace. If "knowledge" is the currency of both the present and future economy, the quality of the output of our education system can only increase in importance. Moreover, the ability to self-educate as well as transform data and information into productive knowledge may be a kind of "academic skill" more critical to both individual and national success than those particular skills and knowledge represented by current national standards. There remain innumerable questions about why our education system is as it is. The U.S. has faced any number of challenges and produced remarkable results in relatively short periods of time as examples such at the economic response to World War II and the Apollo Project demonstrate.

Given the talents we as a nation have demonstrated, why do thirty years of education performance data reveal such limited improvements? The KERA case study sheds light on the problems with translating a new vision into a working reality but the real answers are yet to be found or perhaps merely assembled into a story that makes sense. Whatever the reasons, the need to ensure that the U. S. education system is serving the needs of the nation in all its manifest diversity is clear. The remainder of this whitepaper will discuss ideas for reforming education beyond, or perhaps within, the new standards-based education paradigm. You may find, as we at TCFIR have, that the story of CAT/L is not about a new idea but the discovery that some answers to the problem of education reform are right there in front of us.

Computer Aided Teaching and Learning

Education is important. The quality of the U.S. education system directly affects the nation's economy through the quality of the workforce available to employers, the demographics of the consumer market, the productive capability of the economy, the pace of innovation, and the relative standing of the U.S. globally. The quality of education determines the life opportunities available to individuals and their ability to exercise their right to "life, liberty and the pursuit of happiness." And the quality of education influences the cultural and social fabric of every nation, even influencing such seemingly unrelated phenomenon like birth rates, marriage demographics, trends in religion and urbanization patterns.

While NCLB represents a major overhaul of our nation's education system, the need to ensure that the education system is doing what it must do demands that we not wait to see whether the states can meet the NCLB targets. New ideas in education are as important and relevant now as they have ever been. CAT/L is a proposal for a framework and an organizing set of principles to guide educational research and development. The idea for CAT/L coalesced around the observation that, despite large sums of money, bright ideas and energetic execution, gains in academic achievement are not what one would hope. We presented evidence from state, national and international sources to support this observation.

The data presented does show that some progress is being made. However, judgment of the success or failure of the education system is something that truly cannot be empirical; it rests with each person's dreams of the future and with the ideals of the nation. Whether the data from the previous part represents success, failure or some mixed state is up to you. TCFIR has taken on education reform because, regardless of the judgment one might pass on the education system, we see many opportunities for improvement. Since the publication of A Nation at Risk in 1983, information technologies have transformed or influenced almost every dimension of our world, including education. But like the main body of reform efforts, information technology has manifested little strong positive effect on academic outcomes.

We cannot point to any trend or set of data and say, "Here it is. Here is where information technology has improved education." Yes, parents and teachers exchange e-mails. Yes, multimedia in the classroom has moved far beyond the film strips and reel-to-reel projectors of days past. Yes, report cards and school assessments are available via the Web. Yes, computer science is part of the education curricula of most high schools. Yes, the Internet has placed within reach of everyone with a computer such vast quantities of information that it is almost unusable. And, yes, there have been innumerable developments in learning and teaching technology, from the LOGO programming language to computer-based learning programmes and Blackboard to online classes.

But despite these developments, information technology has not yet penetrated the core problem of helping students to learn better and teachers to teach better. What is the main effect of information technology? Information technology is an enabler and facilitator of human capability, like a lever that allows a person to move loads far heavier than muscle power alone. Information technology manipulates data but data only becomes information and knowledge in the human context and by processes of the human mind.

One example of the magnification of human capability can be found in an explanation of the long period of economic growth through the 1990s: information technology enabled dramatic increases in worker productivity as businesses found ways of exploiting the capability of personal computers and networks. In other vein, mathematical simulations of climate processes could be produced with pencil and paper, as human Computors once did in calculating books of navigation tables, but computers make it possible to view the results in hours or days rather than centuries and in charts, graphs and visual simulations far more useful than tables of raw data. Information technology in the context of education has already enabled many things not previously possible.

For example, the principle of accountability within NCLB is made practicable by information technology because data can be shared, aggregated and analysed in ways not possible twenty years ago. But information technology must do more for education.

Bounding the entire CAT/L framework is a question: how can information technology and Internet-enabled technology in particular, improve education? This is not a new question but it is a question that remains unanswered. Within the framework is an observation: hundreds of technologies, insights from research, methodologies and processes already exist that address parts of the problem of improving education. Binding the framework together is a premise: these existing pieces of technology, research, methodologies and processes can be integrated by organizing principles and additional research and development to create something new, unique and powerful in education reform. This is one meaning of the name "CAT/L". Technology, then, is the foundation of CAT/L and an Archimedes Lever by which to move the world of learning.

The remainder of the CAT/L framework can be described in a series of principles:

- The education process must become learner-centered.
- Assessment - diagnostic, formative and summative - must be improved and deeply integrated into the learning and teaching process.
- National and state academic standards must be met or exceeded.
- Ethnic academic achievement"gaps" must be addressed and eliminated.
- Learning must become more active.
- The formation of life-long learning behaviours must be facilitated.
- Education reform must be guided by empiricism.
- Well-designed, technology-enabled education reform will be self-improving, selfreforming and self-documenting.

- Teaching and learning content must be of the highest possible quality, current and relevant.
- Proven pedagogical methodologies and the best research from all fields with a bearing on learning and teaching must be integrated into education.
- The needs of all stakeholders must be served.
- Reform must also address the need to improve the formation and achievement of vocational goals by students.
- Where minimum standards exist, the goal must be near-universal mastery rather than a standard distribution of achievement.

These principles, written as imperatives, sound like inflexible goals but they are in fact standards by which to assess the various activities within the CAT/L framework. Each principle is drawn from research but also reflects a consensus opinion within TCFIR. In addition, these principles represent hypotheses and, as such, are subject to falsification by scientific method and by consensus of the larger scientific and educational community. In total, the goal implied by the principles of CAT/L is ambitious and perhaps even unattainable. But there are too many examples of great successes arising from impossible-sounding ideas to be deterred from attempting an ambitious proposal.

Even partial success has the potential to significantly advance education. For example, what is the benefit to education and to the U.S. if technology and methodologies could be produced that resulted in near-universal mastery of minimum standards? What would be the outcome of identifying to a high degree of "certainty" those pedagogical methods that produce the best results? At present, one can only guess at an answer to either question.

That is why CAT/L is called a framework based on questions and principles. Clearly, the job of education is complicated and difficult if for no other reason than the fact that we are only just beginning to understand the human mind. The problem of reforming education is even more complicated and difficult because, in addition to not clearly understanding the object on which the processes of education operate, there is the system of education made up of people in roles and social structures that perform and manage education and the consequences of getting it wrong.

To paraphrase Donald Rumsfeld, we do not know what we do not know. Therefore, CAT/L is also a journey of discovery. As briefly discussed in the preface, education is connected to all dimensions of society and the individual. Therefore, the methods and perspectives used to develop CAT/L will come from any and all disciplines. We submit that education reform must be interdisciplinary to be both effective and practicable.

However, interdisciplinary research and development, in and of itself, will present a great challenge because the sum of human knowledge is large and by necessity has become fractured into specialties and sub-specialities. How an epistemologist might relate to computer programmer and a political scientist on a problem of education is a serious question and a significant challenge. But if information technology is to be crafted into a form that promotes the acquisition of human knowledge and implemented within a large bureaucracy, such a meeting of minds will have to happen and be productive.

The Principles of the CAT/L Framework in Brief

Assessment against national academic standards is a major principle of NCLB and recent state-level reform efforts. Assessment methods are generally categorized just as to three types: diagnostic, formative and summative. The summative method of assessment is the most common type and includes quizzes, exams, writing assignments, reports, etc. but also include the various NAEP instruments and high school exit exams. The distinguishing characteristic of summative assessment is that it is typically used after a period of learning and represents a judgment about the learning that has taken place.

In practice, each judgment is represented by a grade with each grade over a period of time, i.e. a semester, trimester, school year or career aggregated into a summation of each student's relative mastery of the body of learning. Within the current education paradigm, summative assessment is usually the end of the learning process for a unit or body of material; the teacher moves to the next unit in the lesson plan and students remain at whatever level of achievement they have demonstrated in the assessment.

In contrast, formative assessment may employ similar instruments as the summative method but is used during the instructional process to provide feedback to both the teacher and learner. Information from formative assessment guides further learning and teaching activities and, in an ideal education environment, provides a basis for guiding all students to a uniform mastery of each unit of learning. Diagnostic assessment has a less concrete definition than the summative and formative types. In this whitepaper, diagnostic assessment has two meanings: first, a method of assessment used prior to a unit or body of learning to determine what preexisting knowledge and skills students have, and two, assessment used to determine traits of each student that influence learning.

In the first meaning, "diagnostic assessment" includes both summative and formative assessment methods as each method, in a dynamic context, provides information useful to subsequent teaching and learning. A subtype of assessment important to CAT/L is certainty assessment. Certainty assessment adds a dimension of data in summative and formative assessment methods by including student self-assessment of the certainty with which they have answered a question. Certainty assessment is currently employed by the company Knowledge Factor in their proprietary industrial learning systems and in the United Kingdom by researchers and universities in a variety of settings.

In formative assessment, certainty ratings allow both the learners and teachers to understand qualitatively how well discrete units of knowledge and skill are apprehended. In combination with formative assessment processes overall, certainty assessment facilitates precise adjustments to pedagogy, focused remediation of material not mastered and an empirical base to guide the pace and direction of a lesson plan.

In addition, certainty-informed diagnostic, formative and summative assessment provides detailed information about "delta" in the student that may be used to formulate a new "grading system" if one is deemed necessary. The importance of improving and deeply integrating assessment into the teaching and learning process cannot be overstated. Simply stated, one cannot understand, control or improve what one cannot measure.

Improved assessment methods bring empiricism into the classroom and provide a precision, efficiency and productivity to the education system that has previously been missing. In a psycho-social context, improved assessment methods facilitate learnercentered practices by promoting a partnership between teacher and learner; reduce or eliminate the need for "high stakes" assessment like high school exit exams because a student's progress through the body of education standards will be well known; potentially reduced stresses associated with assessment, *i.e.* "test anxiety", by a) increasing student preparedness for summative exams and b) habituating students to assessment as a means of learning; and possibly reduce or eliminate social stigmas attached to grades and academic achievement as the meaning of assessment is different in this approach.

Finally, because detailed information along multiple dimensions will be developed by an improved assessment system, other practical uses for this information will be discovered. One potential use is helping students with the process of identifying and accomplishing academic goals related to vocation.

National and State Academic Standards Must be met or Exceeded: Any reform effort that ignores or seeks to bypass the current paradigm of standardsbased education is doomed to failure because NCLB is the law of the land. A common criticism of education standards coupled with accountability practices is the potential to narrow the scope of education to only that which produces the best assessment results. Regardless of this and other criticisms, standards are a reality that must be embraced. The principles of CAT/L–especially improved assessment, learner-centered practices and appropriate uses of technology–are, as a body, intended to produce reform that helps the U. S. education system meet the targets set by NCLB. But CAT/L in its full prospective manifestation has the potential to move academic achievement beyond national standards and address latent needs for adaptive life-long-learning, workplace achievement, critical thinking, metacognitive skills useful in a "knowledge-based" economy and relative national economic competitiveness.

Ethnic Academic Achievement "Gaps" must be Addressed and Eliminated: The persistent quality of the disparity in academic and socio-economic achievement between ethnic groups in the U. S. despite numerous, targeted reform efforts suggests that there is a "structural" problem with the education system. Learner-centered practices show promise in addressing the ethnic achievement gap but multiple lines of new research and development may also be required. In addition, the technological base of CAT/L leads to a problem in addressing the ethnic achievement gap: the prevalence of computers and high-speed Internet connections are lower in non-White ethnic groups and in low socio-economic status schools and districts. Regardless of the challenges, demographic shifts in the U. S. population make addressing the ethnic achievement gap more important than ever.

Learning Must Become More Active

There are numerous justifications for this principle but one will suffice to communicate the point.

If we the accept the premises that:

- Conditions are changing rapidly in technology, the workplace, within the economy, etc.,

- The pace of change is unlikely to abate,
- Current concepts such as the "knowledge-based economy" and "knowledge worker" are valid and tokens of actual processes in action,
- A major adaptive trait of individuals in these conditions is the ability to self-educate, then "active learning" is a concept that has great value both in the posteducation setting and in the education setting as "simulation" of the environment that students will face upon entering the knowledge-driven workplace.

In this practical justification, active learning is equated with self-education although the two are not synonyms: active learning most aptly applies to structured or guided learning settings while self-learning is a superset of active learning but descriptive of any self-directed learning in any learning context. In this perspective, active learning can be offered as an ideal of the learner "self-educating" even within the context of the classroom. Active learning, however, requires some knowledge of both learning and thinking on the part of the learner, areas addressed in the cognitive and metacognitive domain of the Learner-Centered Psychological Principles. Consider for a moment your own experience within the education system and then answer these questions. Were you ever taught strategies for thinking or problem solving? Were you ever given explicit methods for effective study or research? Were you ever presented with opportunities to connect pieces of the knowledge or skills you acquired to other pieces or explore applications of your knowledge in a variety of contexts? Most likely, your answer is "no" to all three questions and yet these active learning skills are fundamental to all effective learning and to the application of academic experience to life.

The Formation of Life-long Learning Behaviours must be Facilitated : Life-long learning is a generative and creative process driven by curiosity and exemplifying the power of intentionality in learning. In the context of our change-driven world, lifelong learning is adaptive. The principle of life-long learning shares the same set of justifications as active learning. A large body of research exists on life-long learning but anecdotal evidence alone from discussions with educators suggests that life-long learning behaviour is a relatively rare trait. This principle is included in the CAT/L framework as a guide to additional research to identify those factors in the current education system that discourage the acquisition of life-long learning behaviours and discover those factors which promote life-long learning.

Education Reform must be Guided by Empiricism: Improving and integrating assessment into the education process is but one part of promoting empiricism in education. The Learner-Centered Model and LCP represent two other dimensions of empirical education reform. Simply stated, empiricism in education reform means replacing all ineffective parts of the education system with new parts validated by research as effective.

To do this, interdisciplinary research is demanded because the system of education, its parts and its connectedness to the larger social system contain manifold dimensions that influence the outcome of the current education system and any reformed system one might envision. No one scientific discipline contains a perspective which encompasses all dimensions of the system of education. Empiricism is a core value of the CAT/L framework.

Well-designed, Technology-enabled Education Reform will be Self-improving, Self-reforming and self-Documenting: "Education reform" is a never-ending process driven, in part, by culture change. Just as the principles of active learning and life-long learning have value to the individual in a dynamic society, so to can an education system benefit from technology that facilitates adaptation to the needs of the society it serves. Improved assessment systems provide information that both teacher and learner can use to "reform" the learning process on the micro-scale. Aggregated assessment data tied to students, student populations, learning content, lesson plans, curricula, schools, districts, states and the national education system can produce new and continuously updated data to assess and reform the system at the mezzo- and macro-scale. Internet-enabled communities and communication can promote collaboration and further reform development. Technology can facilitate the shift from episodic reform to continuous, organic reform much in the same way that "continuous quality improvement" has influenced industry.

Teaching and Learning Content must be of the Highest Possible Quality, Current and Relevant: The Internet is a strange phenomenon, both over-hyped and under-exploited. However, it can be said without hyperbole that within the Internet is more information about everything than any one person can apprehend. Separating canon from apocrypha is another matter. Harnessing the Internet as a dynamic source of educational content is a principle focus of CAT/L.

The intersection between the topics of educational content and the Internet is vast, therefore only a few points will be developed here. Education content across the entire U.S. education system suffers from the same heterogeneity as funding, infrastructure, student and teacher demographics, etc. Well-funded school systems have better content than "poor" systems. Internet content has the potential to replace or supplement traditional content thereby equalizing content quality across the nation. Content quality is further improved by access to new material as it is developed and to alternative content that can facilitate learner-centered practices. Integration of improved assessment practices with Internet-enable technologies results in the potential to assess, learn and teach simultaneously and even non-invasively. Internet-enabled teaching technologies have the potential to streamline administrative tasks such as lesson planning by providing teachers with access to existing, successful lesson plans and support materials developed by Internet communities. TCFIR is currently preparing a report on educational content from the perspective of the CAT/L framework.

Proven Pedagogical Methodologies and the best Research from all Fields with a Bearing on Learning and Teaching must be Integrated into Education: This principle is related to the principle of empirical education reform but is presented separately because it speaks specifically to empiricism applied to the dynamics of teaching and learning. Even a relatively brief survey of education literature reveals a great diversity in the theories and practices of education and yet the core model if U.S. education has remained relatively unchanged for several decades.

There is a disjunction between educational theory and research, and the execution of education within the current system; traditions persist despite superior alternatives and numerous reform efforts. A new survey of the body of educational theory and research–in combination with a review of related, interdisciplinary material–is being undertaken by TCFIR with the goal of presenting this work to scholars for critical analysis.

It is expected that certain patterns and insights will emerge to guide a subsequent process of research and development with the ultimate goal of identifying additional "best practices". The needs of all stakeholders must be served. This principle is an extraordinary challenge but one that is necessary if reform is to succeed. TCFIR is developing a paper which will elaborate on this principle.

Reform must also Address the need to Improve the Formation and Achievement of Vocational Goals by Students: Extraordinary pressure is placed on today's students to make good decisions about education and vocation. While a high school diploma once served as the base for a spectrum of career choices that could produce a middle-class lifestyle, a four-year college degree has become the new de facto standard. Seventeen years of education represents an enormous investment just to meet a minimum standard of qualification to enter the workforce beyond minimum wage. The consequences of making a bad decision are considerable. While meeting the targets of NCLB should improve the academic skills that are the base of any vocational choice, NCLB mandates no reform that would improve the assistance that students might receive in identifying their interests and aptitudes and translating these insights into clear vocational goals.

Trends, such one-third of engineering students changing their majors before graduating, support the need to address this issue. In combination, the principles of CAT/L should address many important dimensions of vocational selection. The enriched feedback of improved assessment methods may provide new kinds of information to students, parents, teachers and guidance councilors. Learner-centered education and active learning practices should promote engagement and the intimacy of contact with all subjects thereby improving learners' understanding of their choices. However, additional research and development will be conducted by TCFIR to explore this issue.

Where Minimum Standards exist, the Goal must be near-Universal Mastery rather than a Standard Distribution of Achievement: The meaning of this principle rests with an understanding of "minimum". Minimum academic standards imply a body of knowledge and skills that everyone must have to function and participate in society. One can debate the meaning of an ill-defined phrase like "function and participate in" as it might relate to individual fulfillment or national need and the responsibility society has, through the education system, to serve this need. However, the intent of such a statement is clear: education must serve students so that they are at least minimally prepared for life.

This is the implication of "no child left behind". Earlier, we argued that the expectancy of a standard distribution of academic achievement in any population of students represented a tacit acceptance of failure but this is only true based on the relationship between the mean of the distribution relative to a minimum standard, and in the context of a mass-education paradigm where there is little done to address the needs of students on the left side of the distribution. Human variation ensures that no population of students can be made uniform in their academic accomplishments. Reforming education so that all students have certain basic academic abilities, while also enabling students to reach the highest level of their native abilities beyond those minimum standards, is possible in the context of a system which emphasizes the development of students over student populations.

Conclusion

CAT/L is an enormous and ambitious framework for educational reform. In this whitepaper, we have presented a summary of the current education system and only the broadest of outlines of CAT/L as an alternative. Neither topic has been treated to any great level of detail. As the purpose of this whitepaper is to serve as an introduction to both topics–even to those familiar with the education system and reform–the impression developed by its content should suffice. In addition, CAT/L is a work-in-progress awaiting new research, insights, contributions and criticisms.

Therefore, more about specific dimensions of the framework will be presented in topic-specific papers. Two points should stand out above all others: the current education system is not a failure nor is it serving the needs of our nation, and CAT/L is not a completely original idea but it is unique in its attempt to organize and develop disparate, existing pieces into a powerful, effective whole. Our efforts depend on a level of interdisciplinary research and development that has rarely been attempted but the need to incorporate the best insights from every field of science that has studied education is clear. Creating a system of education that truly ensures no child is left behind demands that no stone is left unturned.

Chapter 5

Computer: Online Teaching and Learning

Teaching in Online Learning Environments: Overview

What is Online Learning

The term online learning includes a number of computer-assisted instruction methods.

Two parallel processes take place in an online environment:

1. Students become more active, reflective learners.
2. Students and teachers engage in learning through the use of technology and become more familiar with technology by using it.

Online learning is most effective when delivered by teachers experienced in their subject matter. The best way to maintain the connection between online education and the values of traditional education is through ensuring that online learning is "delivered" by teachers, fully qualified and interested in teaching online in a web-based environment.

Approaches to Online Learning

Two approaches to online learning have emerged: synchronous and asynchronous learning. Synchronous learning is instruction and collaboration in "real time" via the Internet.

It typically involves tools, such as:

- Live chat
- Audio and video conferencing

- Data and application sharing
- Shared whiteboard
- Virtual "hand raising"
- Joint viewing of multimedia presentations and online slide shows

Asynchronous learning methods use the time-delayed capabilities of the Internet.

It typically involves tools, such as:

- e-mail
- Threaded discussion
- Newsgroups and bulletin boards
- File attachments

Asynchronous courses are still instructor-facilitated but are not conducted in real time, which means that students and teacher can engage in course-related activities at their convenience rather than during specifically coordinated class sessions.

In asynchronous courses, learning does not need to be scheduled in the same way as synchronous learning, allowing students and instructors the benefits of anytime, anywhere learning.

Course Software

Rather than creating your online course from scratch, a number of software programmes are now available that make it easy to develop an online course. These programmes include features such as threaded discussions and document sharing and pre-designed design layouts to make the course design process easier. Check with the campus technology specialists to learn more about the preferred software for online learning in your department.

Advantages of Learning Online

Online learning offers a variety of educational opportunities:

- *Student-centered learning*: The variety of online tools draw on individual learning styles and help students become more versatile learners.
- *Collaborative learning*: Online group work allows students to become more active participants in the learning process. Contributing input requires that students comprehend what is being discussed, organize their thinking coherently, and express that thinking with carefully

- *Easy access to global resources*: Students can easily access online databases and subject experts in the online classroom.
- *Experiential learning through multimedia presentations*: New technologies can be used to engage and motivate students. Technology can also be used to support students in their learning activities.
- *Accessible for non-traditional students*: Online delivery of programmes and courses makes participation possible for students who experience geographic and time barriers in gaining access to higher education.
- *Draws on student interest in online learning*: Many students are interested in online learning. In a recent survey conducted by the Office of Academic Planning and Assessment at UMass Amherst, more than 50% of students surveyed said that they were "very interested" or "somewhat interested" in taking an online course.

Advantages of Teaching Online

Teaching online courses can:

- *Offer the opportunity to think about teaching in new ways*: Online teaching can allow you to experiment with techniques only available in online environments, such as threaded discussions and webliographies.
- *Provide ideas and techniques to implement in traditional courses*: Online e-mail discussions, a frequently-used practice in online learning, can be incorporated into traditional courses to facilitate group work. Other techniques, such as web-based course calendars and sample papers posted on the Internet can easily be incorporated into a traditional course.
- *Expand the reach of the curriculum*: Online teaching can expand existing curriculum to students on a regional, national, and international level.
- *Professional satisfaction*: Teaching online can be an enormously rewarding experience for teachers. Teachers often cite the diversity of students in online courses as one of the most rewarding aspects of teaching online.
- *Instructor convenience*: Teaching online can offer teachers conveniences not available in traditional classroom settings; for example, at-home office hours and flexible work schedules.

Challenges of Teaching Online

According to a recent American Federation of Teachers report on distance learning, faculty must be prepared to meet the special requirements of teaching at a distance.

Some of the challenges for instructors of teaching online include:

- Familiarity with the online environment
- Capacity to use the medium to its advantage
- Being available to students on an extended basis electronically
- Providing quick responses and feedback to students

Yet, the proponents of online learning argue that these obstacles can be overcome by employing such techniques as the following:

- *Become familiar with the technology used in your online course*: Long before your course starts, become familiar with the technology used in your online course, including hardware and software, and spend some time exploring their options. An online course requires a high level of computing power and reliable telecommunications infrastructure. Make sure you have access to both.
- *Use the online medium to your advantage*: The online environment is essentially a space for written communication. This is both a limitation and a potential of online learning. Written communication can be more time consuming, but "the ability to sit and think as one composes a question or comment also can raise the quality of discussion." Additionally, shy students who have trouble participating in a classroom discussion often feel more comfortable in an online classroom. Online classrooms can be developed with this fact in mind to take advantage of these considerations.
- *Keep connected with students*: Use the technology of the online environment to help you keep in touch with students. Communicate frequently with students, both individually and as a group. A main part of this focuses on how to connect with students. While keeping connected with students can be a challenge, the online environment offers a number of interesting pedagogical opportunities.

Common Terms

Following are some common terms used in online courses:

- *Lurking*: Reading threaded discussion responses without posting a response. Students who lurk in online courses are like silent students in traditional courses; they listen but do not speak. In online situations where you do not know how many people are "listening," lurking can be problematic if others do not know you are present.

- *Threaded discussion*: An asynchronous discussion. In threaded discussions students may post responses to a prompt at any time. Threaded discussions allow students to work at their own pace, allow the teacher to respond more thoughtfully since all the responses are not posted simultaneously, and are easier to coordinate than expecting all students to be online at the same time.
- *Webliography*: An online bibliography of web-related resources. Often online teachers will use a web-based bibliography to help students identity appropriate Internet resources.

Teaching an Online Course

Preparing to Teach Online

As you plan your online course, it is helpful to remember that in any environment "good teaching is good teaching". Experienced online instructors stress that teaching online is less about the mechanics of distance education and "more about what makes for an effective educational experience, regardless of where or when it is delivered". Many teachers have found the Principles of Good Practice in Undergraduate Education to be a useful framework for thinking about how to enhance student learning in their classes.

Principles of Good Practice in Undergraduate Education:

- Encourages contact between students and faculty, especially contact focused on the academic agenda.
- Develops reciprocity and cooperation among students, *i.e.*, teaching students to work productively with others.
- Encourages active learning, *i.e.*, doing and thinking about the learning process.
- Gives prompt feedback and helps students understand how to respond.
- Emphasizes time on task by providing repeated useful, productive, guided practice.
- Communicates high expectations and encourages students to have high self-expectations.
- Respects diverse talents and ways of learning and engenders respect of intellectual diversity.

An additional good practice that does not appear on this list, but that many experienced online instructors mention as being essential to successful teaching, is:

- Includes a well-organized course, the structure of which is clearly communicated to students.

Use these eight best practices as a framework for thinking about your online course. Of course, it is also important to acknowledge that some aspects of good teaching, such as faculty-student contact and cooperation among students, are particularly challenging to accomplish in an online environment. This recommendations on how to accomplish these goals, despite the complications that may exist.

Preparing Students to Learn Online

Students new to online learning may initially find this kind of learning disorienting without the physical classroom space and guidance from the physical presence of a teacher. Other students may initially misperceive learning online as "easier" than learning in a physical classroom space. In reality, students often find the workload in an online course heavier because they must cover course material on their own and type their discussion comments.

There are a number of suggestions for how to help prepare students for online learning:

- *Clarify computer skills/terminology*:
  - o Provide guidelines that detail the minimum technological requirements needed for the course.
  - o At the beginning of the semester, provide a detailed worksheet with instructions on how to complete the technical tasks required for completing course work. For example, while it may be clear to you how to post a message for many students, such tasks are new. Also, while some students may be familiar with one online environment, do not assume that they are familiar with all online environments. Some examples of information to provide include:
    - — Where to find information online
    - — How to post a message and homework assign-ments
    - — How to access course readings and take online exams
  - o Describe how to seek help immediately when having trouble
  - o Explain online conventions for tone, such as using ALL CAPS for emphasis. Set rules for using abbreviations and emoticons.
  - o Provide a tutorial on computer basics.

a. *Tip*: If you cannot provide a tutorial on computer basics, try working with a local community college to schedule a series of computer orientations. One Online Fellow scheduled five 2-hour computer orientations at a local community college to help her students learn computer basics. The college's IT staff setup 15 computers with updated browsers and word processing. Students learned basic computer operation as well as word processing skills. The final tutorial was dedicated to navigating library databases and the World Wide Web. The collaborative effort helped ensure student success for those students unfamiliar with online learning.

- *Explain the differences in learning online versus learning in a traditional classroom*:
  - o Emphasize the amount of time needed for taking an online class and the importance of working independently. Because all class discussions are written, students must be prepared for the amount of time needed to type their comments. A 3-credit online course can easily require more than six hours of time, especially for students who type slowly.
  - o Emphasize the extensiveness of reading and writing in an online course. Because all class assignments are provided in written format with no opportunity for class questions, teachers detail class assignments thoroughly in online courses. Consequently, students must become careful readers in order to ensure that they understand the assignment.
  - o To help students understand the communication differences of learning online, provide a detailed worksheet with instructions on communication guidelines.

    a. *Tip*: Conventions for Communicating Online to this stage for an example.

    b. *Tip*: One instructor uses the following explanation to clarify to her students the definition of a threaded discussion post.

How much to post, and what makes a "good" post?: These are hard questions to answer because discussions are organic, developing and evolving depending upon what is said by whom. In general, posting only once is not enough to really engage in a discussion. I am expecting probably 3-

6 posts depending upon the amount of time I've allotted for the discussion and how in-depth your posts are. What I expect and hope to see is a dialogue evolving, with give and take, back and forth, questions asked and ideas explored like in a face-to-face class discussion. So as you post be cognizant that you are engaging in a discussion. Do not post long pages of responses—probably a couple of paragraphs at most, sometimes a sentence or two can be effective, especially if you're asking a question.

- *Address students' concerns on cyber-culture anxiety*:
  - o Encourage questions and comments about technology.
  - o Use a survey to assess student technical knowledge at the beginning of the semester.
    - a. *Tip*: The Pre/Post Survey example in this stage for ways to assess student technical knowledge.
- *Clarify expectations*:
  - o Post guidelines for participation on the class homepage. For example, explain to students how many days each week they should login to the course website. In online courses, it is not uncommon to expect students to login every week day.
  - o Give a detailed, conspicuous course outline. Because you must clarify course expectations only in writing, make sure that you give students enough detail to complete class assignments. Even simple assignments like a journal need detailed explanations.
    - a. *Tip*: One instructor uses the following explanation for the weekly journal exercise. She posts this explanation in every unit to remind students weekly of the assignment.

Your journal is the place for you to keep thinking about, wrestling with, exploring the issues we've discussed online. Feel free to add your own day-to-day observations about issues related to our course. Your journal is only read by me. I will never comment on your observations; I only check to If you've completed the assignment. Length: 1-2 paragraphs Due: Every Friday by midnight EST

  - o Set clear expectations with regard to student performance/ activity. Help students understand expectations for the course and encourage them to ask questions. One way to help students understand course expectations is to post examples of model assignments. You can post examples of model assignments from other webpages or upload sample papers. Most online course software programmes allow you to easily upload files, such as MS Word and Excel documents.

- o Remind students frequently of course expectations.
  a. *Tip*: During the semester, one instructor posted reminders to keep students up-to-date with the course material. Following is an example of one such reminder: Have you read your James McBride?: If you haven't started reading The Colour of Water, you better get reading!It's almost Monday and the weekly exercise is due Wednesday. Look forthe threaded discussion posting on Monday morning.Hope you had a great weekend! I look forward to getting your responsepapers on Monday night!
- o Explain the time-frame in which e-mails will be answered. For example, on Monday, Wednesday, and Friday only, or within 2 business days of receipt.
- o Emphasize courtesy to fellow students. Because students can not see verbal or visual clues from other speakers, encourage them to be tactful in their responses or include parenthetical clues for humour or emotion.
  a. *Tip*: One instructor describes courtesy to her students with the following explanation.

Our online discussions will be class discussions, meaning the same respect we would show each other in an actual classroom, as suggested, also show in a virtual classroom. In fact, because the online environment is primarily a verbal environment where we communicate through writing, it lacks the physical and auditory clues that accompany face-to-face discussion, which may lead to more misunderstandings, particularly when a person is using humour. But being polite and respectful does not mean that you can't disagree or question each other's interpretations of our texts. But be sure to do so in a polite way, rather than "I think you're wrong and here's why" write instead, "Sally, I think you are saying Y [paraphrase what person wrote], but I wonder if there isn't another way to look at that same incident. The way I see it, X really happened..." etc.

Teaching and Learning Challenges

Structuring an Online Course

Experienced online instructors and students alike emphasize the need to have a clearly structured and well-planned course when teaching and learning online. Structuring the course effectively means planning the course well in advance of when it is being taught, thinking through the organizational structures and qualities that will help students learn, and understanding that the online environment presents a number of communication challenges.

Course Planning

Designing a course always takes a great deal of time and thought. That is no different with online courses. At the same time, the online environment offers particular obstacles and opportunities for both instructors and students. As you think through the course elements, pay particular attention to the course components that may serve as stumbling blocks to student learning online. One particular tension that emerges is the need to have a clear and organized structure, while allowing flexibility for making adaptations mid-stream.

- Develop your course before the semester begins: Often new faculty discover that developing online courses is time-consuming and that transitioning a successful traditional course to an online setting can be difficult. Experienced online instructors suggest developing your course well in advance and with a clear, concise objectives statement. The better prepared you are, the better your online teaching experience will be.
- Allow flexibility in your course design: Although it is important to make course expectations and due dates clear, it is also important to build in flexibility to your schedule. Building flexibility into your course structure will allow you to compensate for unexpected technological problems as well as give you opportunities to respond to student feedback.

Course Organization

Students in online courses are in particular need of a clear organizational structure. Keep in mind that each student is experiencing the course on his or her own - without the opportunity to turn immediately to a neighbour if confused or unclear about something in the course. In addition, students in online courses do not have the imposed structure of attending class at a consistent time and place each week they do not have the traditional"markers" of handing in papers in class or coming to the classroom to take a test. For all these reasons, it's important to think carefully about how to appropriately organize your course to encourage student participation and facilitate student learning.

- Chunk the syllabus into parts: Divide the course syllabus into discrete segments, organized by topic. Self-contained segments can be used to assess student mastery of that unit before moving forward in the course.
  - o *Tip*: Use an"Assignments" page for course assignments. On that page, outline each assignment in a paragraph, explaining its purpose in helping students, and provide explanations and guidelines for evaluation. The sample course homepage in this stage for an example of organizing your course this way.
  - o *Tip*: Another way to divide the course is by time. One instructor uses the following organization, in which each unit is labeled by week and author, for her literature course. The first two weeks of her course:
    - a. Course Home
      1. Syllabus
      2. Calendar
      3. Lounge
      4. Questions
    - a. *Week* 1: Kyoko Mori
      1. Who is Kyoko Mori?
      2. Reading Notes
      3. Weekly Exercise
      4. Journal
      5. Threaded Discussion
    - a. *Week 2:* Esmeralda Santiago
      1. Who is Esmeralda Santiago?
      2. Reading Notes
      3. Weekly Exercise
      4. Journal
      5. Threaded Discussion
- Break assignments into chunks with "touch points": Because students work at their own pace in an online course, it works best to develop guidelines that require students to come back to

the course website often. Chunking assignments helps students keep up with the work. In addition, use "touch points" at which point students do something–write in a journal, send an e-mail, enter into a discussion–to help chunk course content and give the course more structure.

- o *Tip*: A literature instructor chunked one unit as follows:
    a. *Assignments for their eyes were wathcing god*:
        1. *Background information*: Before you begin to read Their Eyes Were Watching God, please read the background information that I have provided.
        2. Read stages 1-10 and write a two-page, single-spaced reading response that you will put in your Journal on the course homepage journal link. This response will be more informal than an essay, and is due by midnight, July 23.
        3. Finish Reading the book and post at least twice to your group discussion board 8 p.m. July 25.
        4. Respond to your group discussion board several times© by noon, July 26.

- Provide due dates for assignments: Each assignment should have a clear due date and time. In addition, multiple due dates every week keep students on track with course requirements.
- *Provide multiple opportunities for graded activities*: Assess students on writing assignments, standard test formats, and class participation. The online course format offers a number of opportunities for graded written assignments, including threaded discussions, papers, web research, and online exercises. Multiple measurement points will stimulate students to become involved in multiple activities and keep them participating in class.
- *Give credit for participating in online discussions*: Give students credit for the substantive learning that students provide for each other through online discussions. In many online courses, these discussions are essential for advancing the course goals. By assigning credit for participation in online discussions, instructors can deter "lurking," where students listen to the conversation but do not participate.

Communication

In considering how you communicate with students about course goals and your expectations, it is again important to remember that students experience your course on their own and will come to the course with varying levels of technical expertise. Place important information in a variety of places, and repeat it often, in order to enhance the chances that students will pay attention to it.

- *Give students a clear overall understanding of the course structure*: Students need a clear message of the "vision" of the course so provide them a sense of the overall landscape of the course.
    - o *Tip*: Use a Table of Contents layout design to help first time online students understand the structure of the course. The Table of Contents style is similar to printed material. The sample course homepage in this stage for an example of how to provide a sense of the overall landscape of a course.
- *Post course syllabus, policies, expectations, and objectives on the course website*: You will most likely not be available to respond immediately when students e-mail questions regarding assignments or due dates, so posting your syllabus on the course homepage will eliminate confusion.
    - o *Tip*: Students will access the course homepage at any time of the day or night. You can't always be online to answer questions, so make the assignments easy to find and easy to understand.
- *Setup a housekeeping clearinghouse part on your webpage*: To cut down on the number of individual questions, set-up a housekeeping clearinghouse part on your webpage where students can post a question and get answers about general course information. Encourage students to go to this part of the course before asking the instructor.
- *Use printed materials if a student requests*: Have a printed workbook of course syllabus and other critical course information available for students who request printed copies.
    - o *Tip*: For engineering courses with heavy math content, provide detailed lecture notes, solutions, and other course materials in PDF format before the lecture date or online access date. This will allow students to download and print course material in advance.
- *Structure online discussions*: Structure the course to capitalize on the threaded discussion format. Use existing textbook material or website readings for "lecture" and guide students through activities and threaded postings for active learning.

- *Remind students frequently of due dates*: Use a technique like "Nag Notes" to remind students of due dates and other requirements.
    - o *Tip*: One Communication professor uses "nag notes" to remind his students of due dates. For example:
        a. I've posted the topics proposed thus far. Browse to projects/paper No. 1. Reminders:
        b. For Wednesday, Read the Birkerts piece, "Into the Electronic Millennium."
        c. For Monday, Read Postman's and do the IT/HC in the News Discussion Forum assignment.

Creating Community

In an environment where instructors do not necessarily meet students face-to-face and where students may never have an opportunity to meet their peers in a physical classroom, developing a sense of community can be particularly challenging. At the same time, a sense of a community–where students are able to work cooperatively with peers on course material, have the opportunity for positive interaction with the instructor, and where the learning environment is respectful and motivates students to do their best–is key to a positive and successful learning experience.

This part provides a number of solutions for creating community in the online classroom. The Online Fellows are quick to point out, however, that creating community is a challenge, and classroom dynamics must be monitored throughout the semester to ensure that students continue to engage thoughtfully in course content and continue to work together productively.

Student-to-Student Interaction

As the Principles of Good Practice in Undergraduate Education make clear, student learning in any classroom is enhanced when students have the opportunity to connect with each other about their academic work. For the online instructor, facilitating student-to-student interaction is made particularly challenging because students do not naturally have a chance to get to know each other before class or in face-to-face conversations. Therefore, it is important to structure opportunities where students "have" to interact with each other. It is also important, however, that the instructor develop methods for monitoring the success of these interactions. The Online Fellows offer the following recommendations:

- *Limit the size of discussion groups*: Rather than having an entire class talk in one large group, break the class into smaller discussion groups of four or five students. That way, students can get to know each other in a more intimate way.
- *Allow students to post student-to-student communication to get answers to questions*: Encourage students to discuss among themselves. Do not respond to every comment-interject and guide the discussion. Encourage students to introduce themselves to the group at the beginning of the semester.
- *Pair each student with a"buddy" in the course*: The buddy system gives students a source of support in the online classroom. Some instructors match students with varying technological experience. Other instructors prefer to match students who possess similar technological skills. Pair students according to the goals of your course or the assignment.
- *Encourage peer response*: Post student papers online and ask each student to select a partner to critique each other's work. Be sure that students know their paper will be posted.
- *Structure opportunities for personal interaction*: Incorporate opportunities for students to tell you something about themselves in a"student lounge" or meeting place. A"student lounge" can also be a place where students can share with each other, meet each other virtually, and learn more about each other without your presence. The"Student Conference Centre" at the end of this stage for an example of how to set up such a location on your course website.

Faculty-to-student Interaction

The Principles of Good Practice highlight the importance of faculty-student interaction in promoting learning. The online environment is not necessarily conducive to this goal, because neither the instructor nor the student can rely on regular face-to-face interactions to reinforce one's willingness to be helpful and approachable.

Experienced online instructors, however, have identified the following ways to help enhance faculty-student interactions:

- *In your written communication, present yourself as accessible to students*: Students in an online course must feel that you are approachable. Often the demands on teachers are greater in online courses, so it is important to explore the variety of ways you can send a message of availability. One way to bridge the distance between faculty and student is to address students by name. Praise student-initiated contact.

- o *Tip*: To make yourself seem approachable to students, try using a more informal tone. For example,"Today, as you all are well aware, our class officially begins. Please begin working on the assignments for July 15-21. You have a couple of assignments due tonight"

- *Schedule an in-person meeting of the entire class*: If possible, meet with students in person for one session at the beginning of the semester. Meeting in person helps students associate names with faces and can be an effective, timely way to accomplish many of the administrative tasks central to your course.
- *Generate frequent communication*: Students need to have a sense the instructor is really "there," not "missing in action." This means responding in a timely manner to individual questions or issues that are raised in discussion groups. It also means making your presence known by participating in online discussions, giving students regular feedback on their work and their comments, and being flexible enough to make changes to the course mid-stream based on student feedback.
- *Assign discussion group leaders or project team leaders to facilitate group work*: Assigning team leaders is one way to ensure that students receive ample feedback. Make sure that the team leader disseminates information to every member of the team. Part of the responsibility of the team leader should be to report to you frequently on the progress of the team.

Tone

Remember that in the virtual classroom, neither the instructor nor the student has the visual cues of face-to-face communication. This also means students have fewer methods for determining whether their efforts are comparable to those of their peers and for assessing how they are doing in the class. Students will use the cues that are available to help them understand the classroom climate. Therefore, how the instructor shapes the course climate through written comments and the tone of communications to students is particularly important.

- *"Humanize" the course*: Remember that although you are teaching online, you are still teaching real people, so it helps if you and students can put names with faces. Develop a portion of the course website to post pictures and brief bios of students.

- *Avoid general broadcast questions*: An online course is not a collective but many individuals all reading messages separately. So, a broadcast message like "Are you doing the reading?" is hard for a student sitting at his/her own computer to interpret.
- *Consider the tone of your own responses to students*: Attitude comes through in writing. Are you sounding impatient? Supportive? Praise and model appropriate tone.
- *Use private e-mail for sensitive communications*: Use threaded discussions for group conversations. Use private e-mails to comment on individual student contributions and criticism.

Assessing Student Learning

What is Assessment

The word "assessment" has taken on a variety of meanings within higher education.

The term can refer to:

- Standardized measures imposed on institutions as part of increased pressure for external accountability,
- The process faculty use to grade students' course assignments, or
- Activities designed to collect information on the success of a course, a programme, or a university curriculum.

The suggestions in this stage focus on the latter two components of assessment:

1. Testing/evaluating student performance and providing feedback to students for grading purposes
2. *Assessing whether the course itself is "working" for student learning*: what is going well, what isn't, and how do you know?

This second definition of assessment–determining what's "working" in the classroom– is particularly important in the early stages of innovative course design because assessment makes it possible to:

- Make informed improvements to current practices
- Document success to share with funding agencies, department chairs, etc.

At its best, assessment should be valuable to the teaching/learning process and not another add-on or "make work" of little use to instructors. In fact, assessment activities can be helpful in promoting all of the Principles of Good Practice.

Evaluating Student Performance for Grading Purposes

In assessing online learning, it is important to create a "mix" of assignments that cover the multiple dimensions of learning that online courses can employ. Traditional tests become a smaller part of the grade as you move towards encouraging student interaction on group projects and other activities.

Different forms of assessment include:

- End of semester paper
- Weekly tests
- Group projects
- Case study analysis
- Journals
- Reading responses
- Chatroom responses
- Threaded discussions participation

Communicate Expectations

As was suggested in the previous stage, students in online courses are in particular need of clear information about course requirements and instructor expectations. Therefore, develop specific grading guidelines for course assignments and activities ahead of time so students know in advance what is expected of them.

For example, articulate what are appropriate responses to questions in online discussions, what is a substantive answer versus a superficial response, etc. Providing students with specific examples of the kinds of work you are looking for is also helpful.

Keep Track of Student Performance

The gradebook option in online software packages makes it possible to store all information about students' performance in one place. Many also make it possible for students to look up their own progress on assignments.

Give prompt feedback:

- At the start of the semester, clarify the type of feedback you will be giving so students have a clearer sense of what to expect from you.
- Students want feedback on assignments, but it is often difficult to provide much feedback when you use a number of varied assignments throughout the semester. One instructor uses system to provide a quick response to students.

- A number of grade book features have a comment part where the instructor can give specific feedback to a student on an assignment that can only be seen by the instructor and that student.

Design effective tests:

- Be clear from the start about what is allowed and what is not permitted when students take a test online.
- Because it is difficult to ensure that students taking an online exam are not using their books, some faculty encourage open book exams but place a time limit on how long students have to complete the test. These instructors believe that if a student knows where to go in the text book to get the information they need in a timely fashion then that student has clearly done the reading, and the issue of memorizing the information is less important. Some online course software allows you to limit the time that students may view test questions and post test answers.
- Unlike many traditional classes where students never see their completed exams after they hand them in, students in online courses can usually go back and look at the exam questions at a later date. While this can be a useful learning tool for students, it can lead to additional questions from students about exam content or the wording of a question.

Encourage active learning:

- Help students become more reflective learners by asking them to set their goals for the course at the beginning of the semester. At the end of the course, ask them to return to their goals to reflect upon what they've accomplished.
- The majority of students focus their academic effort on those elements of the course that will affect their grade in the course. Be sure that your grading policies reinforce the activities and assignments you value and that you take advantage of learning activities that are particularly suited for an online course. For example, if you want students to meaningfully participate in online discussions, be sure to include participation as part of the grading scheme.

Evaluate participation in threaded discussions:

- Require students to participate in specific numbers of threaded discussions.
- Have interactive learning activities account for a high percentage of the course grade.

- Identify the qualities you look for in discussions and grade students according to those criteria.

Assessing Whether the Course is "Working"

Assessing whether your course is "working" provides feedback to understand what is useful to students. Josh Bersin in "Measuring E-Learning Effectiveness: A Five-Step Programme for Success" offers a helpful framework for thinking about the kinds of information you can use to determine your course's success. We also provide some specific assessment techniques.

Five Steps to Measure Effectiveness

- *Enrollment*: Is the audience showing up? If students are not enrolling in your course, then they might not know about the course or do not know how to enroll in the course. If the course is an elective course, the course may be named poorly or not located correctly in the catalog.
- *Activity*: Are they making progress? Typically, if the content is appropriate for the audience, students will progress at a reasonable rate. You may find that students move quickly and then stop at a particular point. Such information is valuable to help you assess the usability, relevance and performance of the course content.
    - o *Tip*: Use minute papers or muddiest point exercises to provide feedback. Minute papers and muddiest point exercises work even better in an online environment because students can share them with each other, so students see what other students are thinking about.
- *Completion*: Did they finish? Students who truly complete the course can provide valuable feedback. However, many course software will "flag" a student "complete" even if that student has not completed all the course assignments. Make sure that you can accurately track which students have completed all the course work.
- *Scores*: How well did a student score? In online learning environments, you often can not gauge why a student has scored highly on a quiz or assignment. Did they really learn the material or copy from someone else? Multiple assessments will allow you measure incremental progress towards the final learning goal, so you can measure what exactly a student scored well on and where they have fallen short.

- *Feedback/Surveys*: Did they like it? Feedback is a vital part of online learning. Regular feedback will provide you important details about the course content, assessments, and technology.
    - o *Tip*: Collect mid-semester feedback and alter the course according to student suggestions. ÿ Tip: Survey students at the end of class about their progress. What worked in the course and what didn't?

Competencies for Online Teaching

Information technology is changing the way people live and learn. Not surprisingly information technology is also transformation the nature of teaching. These remarks provide a framework for thinking about such changes and exploring work in progress that is relevant to the development of competencies specific to teaching online.

Competence, Competencies and Certification

Competence refers to a state of being well qualified to perform an activity, task or job function. When a person is competent to do something, he or she has achieved a state of competence that is recognizable and verifiable to a particular community of practitioners. A competency, then, refers to the way that a state of competence can be demonstrated to the relevant community. According to the International Board of Standards for Training, Performance and Instruction, a competency involves a related set of knowledge, skills and attitudes that enable a person to effectively perform the activities of a given occupation or function in such a way that meets or exceeds the standards expected in a particular profession or work setting.

The structure and assessment of competencies may differ from one community of practice to another and even within a community. To facilitate a common understanding of competencies in the context of online and distributed learning some specifications have been elaborated. Typically, a competency is divided into specific indicators describing the requisite knowledge, skills, attitudes and context of performance.

There are different ways to validate that a person has demonstrated the relevant competencies. One of them is through a certification process. Teacher certification is a common practice, and the notion of teacher competencies is fairly well established. However, competencies are generally associated with highly formalized professional activities and not applied to ill-defined tasks. Ill-defined tasks certainly include many forms of teaching. This narrow view of competence runs counter to common sense and professional practice, but brings into attention the mainstream approach to elaboration of teacher competencies where it is essential to clearly identify the conditions of teaching. The delivery environment is a particularly relevant condition to identify competencies for online teaching.

Online and Classroom Teaching

Information technology can be integrated into both online and classroom settings, but the interaction between these technologies and new approaches to learning and instruction may vary. The range of activities available in online settings and the multiple conditions of time in which they take place are evidence that the technology demands placed on online teachers are somewhat more significant than those associated with classroom teachers. Much of what has already been published with regard to online teaching has focused on technical skills and requirements of successfully moderating and facilitating online discussions and chat sessions. This body of literature suggests that becoming an effective online moderator requires training and that there are competencies unique to online environments. In online asynchronous discussions, the moderator's competencies involve allowing learners time for reflection, keeping discussions alive and on a productive path, and archiving and organizing discussions to be used in subsequent sessions.

In online synchronous discussions, the moderator must establish ground rules for discussion, animate interactions with minimal instructor intervention, sense how online text messages may appear to distant learners, and be aware of cultural differences. How are these competencies unique to online teaching? At the applied level, animating discussions, displaying cultural sensitivity and so on, apply to all teachers. At the environment level, however, the ways in which a teacher demonstrates such competence is quite different, which suggests that there are competencies unique to online settings. According to Belisle and Linard the use of IT in teaching calls for additional competencies adapted to new roles and circumstances. Teaching competencies and online teaching competencies have generally been considered separately. However, efforts to interrelate the two are being undertaken by IBSTPI in association with the research centre for Télé-université, Université de Québec.

Implications of Competencies for Online Teaching

The current interest in competencies for online teaching is coming from business and industry, primarily with regard to technical training and professional development courses offered in online settings. It is quite likely that some of the interest in competencies for online teaching is a result of hastily-crafted online courses and inadequate preparation of online facilitators. Clearly technology offers the potential to create and implement highly engaging and effective online environments to support a wide variety of learning goals. It is also quite clear that our capacity to make effective use of information technology in educational settings is impaired by inadequate preparation of teachers and by a shortage of properly trained instructional designers and educational support personnel.

The development of competencies for online teaching should lead to the associated development of training for online teachers and to the certification of online teachers. To develop competencies for online teachers is not without challenge. Competencies are dynamic in nature, and they largely depend on the relevant social context. The constant transformation of IT makes the development of competencies for online teachers a continuous process and demands continuing professional preparation and training for online teachers. Such endeavors will improve our ability to make effective use of technology in learning and instruction.

Quality Standards in Online Teaching and Learning

The challenges associated with online teaching and learning demand new approaches to quality assurance beyond the framework within which higher education institutions currently operate. As Taylor and Richardson assert, there is a need for quality assurance systems which consider "the standard of online information", and at the same time, support academics in the development of high quality online resources.

This chapter presents an approach developed by the University of South Australia, which addresses both these aspects of quality - providing the standards by which online courses are judged, and supporting academics as they develop their own scholarship of teaching in the area of online learning. The Boyer notion of scholarship is a framework for considering academic work that can be applied to online teaching and learning within universities. Boyer identified four scholarships - discovery, teaching and learning, integration and application. His approach is predicated on an understanding of the communal basis of all scholarly activity: that scholarship by its very nature is a public rather than private activity; that it is open to critique and evaluation by others; and that a field of study is progressed through the scholarly activity of building new ideas which are then open to the same processes of public scrutiny.

All of the scholarships are exposed to the same rigorous approaches of peer review as a way of gaining quality, transparency and accountability. Within this framework the scholarship of teaching and learning has emerged as a major theme in the higher education sector. Central to this notion of the scholarship of teaching and learning is that of the 'learning community' - the recognition of the value of relationships and practices that occur in and through the work practices of staff. One way to support and stimulate this kind of collegial activity is to provide structured opportunities for discussion and reflection through a checklist of agreed good practice. Taylor and Richardson advocate the application of this approach to the design and construction of information and commu-nication technology based teaching resources, arguing that independent peer review requires "...the development of an explicit and shared understanding of the scholarship underlying the design and development of these resources".

Such shared understanding, according to Taylor and Richardson can also form the basis for validating the quality of the resources. This paper describes the development of a checklist and supporting website, in which shared understanding about the scholarship of teaching and learning in resources developed for online delivery is made explicit.

The principles underlying the development of this approach are as follows:

- The criteria for the standards of development have been gathered from the full range of relevant academic literature surrounding online teaching and learning. This affirms the work of academics in the area and provides it in a highly practical form which is accessible to a broadly-based audience.
- The approach locates responsibility for the quality of teaching and learning with the academic staff responsible. Staff can use the items to guide the development or redevelopment of their own courses through reflective processes.
- The instrument and its associated website provide an opportunity for just-in-time academic staff development by providing the accepted standards, information about how to meet these and examples of how others have done this.
- The instrument provides a framework to involve other academics in the process of peer review.
- The website is designed to provide a model of best practice, and has been validated using the W3C Mark-up Validation Service, and the W3C CSS Validation service, and complies with W3C Web Content Accessibility Guidelines 1.0.

Review of Other Instruments

In order to pursue this approach, the authors reviewed a range of instruments available through the Internet. Several generic descriptors for online course development and evaluation were identified. A link to the analysis of these instruments is available from URL: resources/online-eval/. Since online teaching and learning is still a developing area of academic activity within universities, and many staff engaged in online approaches have limited expertise, the authors were interested in identifying instruments that provided an educative and explanatory dimension which supported the evaluative function. In effect, this required the instrument to be both comprehensive in scope and specific in detail. A review of the instruments available identified several problematic issues. First, several had been developed to address particular aspects of course development and were partial in their scope rather than comprehensive. Second, many of them were very general, open-ended instruments. Although there may be some justification for this in terms of providing a generic framework, these instruments make considerable assumptions about the level of expertise of those involved in the processes of online teaching and learning. Third, some instruments were found to be comprehensive in their scope, but unnecessarily complex because the instrument and supporting online materials were not integrated.

Finally, most of the online instruments were found to be inaccessible for users with disabilities. The authors noted features in some approaches that were consistent with the objectives of the proposed checklist of agreed good practice. Of particular note is the Michigan Virtual University's Standards for Quality Online Courses and the accompanying Excel-based Course Evaluator tool. The standards addressed in the MVU instrument include several criteria proposed for a checklist of agreed good practice including; instructional design, accessibility, usability and technology. However, the authors were concerned about the complexity of this instrument, and in particular, the lack of a seamless integration between the Excel tool and the supporting online material. Furthermore, the authors contend that aspects relating to accessibility and usability need to be embedded within criteria relating to instructional design, interface design, use of media and technological issues, rather than treated as separate considerations.

Design and Development

Since the instruments reviewed failed to adequately address all of the needs that the authors had identified as important characteristics of a checklist of agreed good practice, it was necessary to develop a new review tool designed to meet those needs. In doing so, the authors recognized the need to build on the experience gained from the review process which had indicated some consistency in the priority placed on certain criteria.

For example, Michigan Virtual University's standards for quality online courses, the peer review proforma developed by the Griffith Institute for Higher Education, the Electronic Learning Institute's criteria and standards used in evaluating Web-based instruction and delivery guidelines, and Lyn Knowitall's expert review checklist all consider instructional design issues, interface design and/or appropriate use of media and technological issues.

The proactive evaluation model proposed by Sims also places importance on criteria relating to instructional design, interface design and elements of content utility, including the accessibility of the content. Similarly, the MVU standards consider accessibility issues, using the W3C Web Content Accessibility Guidelines 1.0 Priority 1 criteria as its benchmark. This review of the literature and available evaluation approaches informed the authors' decision to structure the review tool and associated website around the following areas of consideration:

- Instructional design
- Interface design
- The use of multimedia to engage learners
- The technical aspects of interactive educational multimedia.

The authors opted to embed criteria relating to inclusivity in items associated with all four areas of consideration, since issues such as accessibility impact on the instructional design, usabilitity, use of media and technical functionality of online course materials. The review of instruments also identified a range of different approaches employed to measure the extent to which the various items listed under these major areas of consideration meet the stated criteria.

These approaches include the complex quantitative rating system delivered via an Excel spreadsheet in the MVU's evaluator, simple yes/no checklist formats utilised in Electronic Learning Institute's criteria and standards used in evaluating Web-based instruction and delivery guidelines, open-ended qualitative questionnaire formats employed in the Southern Regional Education Board's criteria for evaluating Web sites, and the CIDOC Multimedia Working Group's multimedia evaluation criteria, and quantitave measures using a rating scale approach with provision for qualitiative responses to open-ended questions, as exemplified in the Griffith Institute for Higher Education's peer review proforma. Based on this analysis, the authors decided to adopt a combined approach, employing a 5-point Likert scale to and a free form text area for comments.

This approach was considered to be appropriate for the design and development of a checklist of agreed good practice, since a combination of quantitative and qualitative measures will most likely yield comprehensive results. In developing this tool, the authors acknowledge that such instruments have inherent limitations, since as Owston observed"....no single model or framework is likely going to satisfactorily capture the complexity of pedagogical, technical, organizational, and institutional issues inherent with Web-based learning".

However, the tool is not intended to be used in isolation from other academic practices. It will be most valuable when it is part of a wider framework of course and programme development and evaluation or established peer review processes. To a very significant extent the intention of the review tool is to generate scholarly discourse around online teaching and learning within the rich environment of an academic community. Summative and formative evaluation has been an integral aspect of the development of the review tool from the point where the authors identified the need to develop an instrument within their own institution.

This involved reviewing a range of instruments which were deemed inadequate for the purpose and audience. After much research, a paper version was developed and circulated to a reference group of online enthusiasts and other interested staff. Feedback was incorporated into a revised version. Using this version, the course materials of a volunteer academic were reviewed and the results were presented to a seminar of staff involved in online teaching and learning. Further revisions were made and a beta version developed. In the next stage of the evaluation, academic staff, professional development staff and students at the University of South Australia will be invited to take part in a trial using the beta version and their feedback incorporated in the final version of the review tool and the online website.

Description of the Review Tool

The preceding part describes the design and development of a review tool comprising a paper-based checklist of agreed good practice and supporting website which provides an educative function, addresses issues relating to inclusivity, and is constructed around four main areas of focus - instructional design, interface design, use of media and technical aspects.

Educative Function

The educative dimension is central to both the just-in-time approach to professional development and approaches which involve more formal educational development. The associated website supports this educative function through the inclusion of features such as hyperlinks to explanations and the relevant literature that are accessed by selecting a"more" link alongside each checklist item, an exemplars part, and additional resources. There is often confusion among reviewers about the difference between general statements about the overall goals and clearly specified objectives. By selecting the"more" link the reviewer can check their understanding of these terms and also learn more about effective techniques for specifying objectives or learning outcomes from the hyperlink references included in the related explanatory screen.

Inclusivity

Items relating to inclusivity such as gender, culture and accessibility have been embedded across the four parts of the instrument. The decision to embed these items rather than to extract them into separate categories was based on the view that essentially the items reflect good teaching and ought to be seen in a more integrated way. Since the supporting website was designed to provide a model of good practice, it has been necessary to ensure that that it too meets W3C Web Content Accessibility Guidelines 1.0.

The accessibility design features incorporated into the design of the site are as follows:

- All pages validate at HTML 4.01 transitional using the W3C MarkUp Validation Service.
- Cascading style sheets are applied for layout and style, and have been validated using the W3C CSS Validation Service.
- Alt text attributes and captions have been applied to all visuals and image maps.
- Redundant text links are provided as footers on each page.
- Care has been taken to ensure that sufficient contrast is provided between foreground and background images, and that content does not rely on colour alone.
- The primary natural language of all Web pages has been specified.
- All tables linearise appropriately.
- Use of scripting languages and reliance on non-html languages has been avoided

- Links open as new pages rather than as new windows.
- All links can be accessed via keyboard control as well as mouse control.
- Menus are grouped logically and skip links are provided.

Areas of Focus

The review tool is constructed around four sets of considerations: instructional design, interface design, the use of multimedia to engage learners, and the technical aspects of interactive educational multimedia. These areas have been developed through consideration of the literature and are described in the following sub-sections.

Instructional Design

Instructional design criteria consider how the strategies and techniques derived from learning theories are applied to the solution of instructional problems in interactive multimedia applications. The importance of pedagogically driven instructional design in the creation of educational multimedia is well documented.

The features considered in instructional design criteria include:

- Whether the learning objectives are clearly stated;
- The appropriateness and accuracy of the content;
- The sequencing of instruction;
- Whether the topics are applied in "real" contexts;
- Assessment strategies and
- The appropriate use of feedback.

Interface Design

Interface design critieria address the quality of the end-user interface and how it affects "... users' perception of the product, what they can do with it and how completely it engages them". Reushle and Sonwalker contend that interface design and related usability factors will have a significant influence on the success of instructional interactive multimedia. As Sonwalker explains, "Users interact with online Web courses through a graphical user interface, so the design of graphic elements, the colour scheme, the type fonts, and navigational elements can all affect how a course is organized and perceived by students". Interface design criteria address all of these usability factors as well as accessibility criteria since as Dey advises "the interface needs to be accessible to as wide an audience as possible".

Use of Media

Effective use of media is a key aspect of educational design. This area of concern considers issues relating to the effective use of interactive multimedia, writing style and accuracy of text and copyright. The term interactive multimedia is used to identify the capacity of digital media to facilitate a range of interactive experiences; the aim being to promote active learner engagement.

Evaluation of the appropriate use of of interactive multimedia considers the ways in which multimedia technologies are integrated into the teaching and learning process to support the learning objectives, promote learner control and "...actively engage learners in creation of knowledge that reflects their comprehension and conception of the information...". Multimedia components such as animations, video and audio also present challenges for users who have disabilities, and those living in locations with restricted bandwidths. The criteria must therefore also consider accessibility features, such as the provision of synchronised captions to avoid precluding certain groups of students from engaging in the learning experience.

Technical Aspects

The technical aspects of interactive multimedia are considered in reviewing educational applications because software and hardware problems can undermine learners' confidence and their ability to form good models of how computers work.

Accroding to Sonwalker, the issues influencing the technological success of online courses include available bandwidth, target system configuration, server capacity, browser software, and database connectivity. In addition to these factors, evaluation of the effectiveness of interactive multimedia applications in online education needs to consider the extent to which the course materials are accessible to all users across different platforms and browsers, if plug-ins are required whether the user is informed and links are provided, whether all hyperlinks are active and the overall robustness of the application.

Chapter 6

The Future of Online Teaching

Institutions of higher education have increasingly embraced online education, and the number of students enrolled in distance programmes is rapidly rising in colleges and universities throughout the United States. In response to these changes in enrollment demands, many states, institutions, and organizations have been working on strategic plans to implement online education.

At the same time, misconceptions and myths related to the difficulty of teaching and learning online, technologies available to support online instruction, the support and compensation needed for high-quality instructors, and the needs of online students create challenges for such vision statements and planning documents. In part, this confusion swells as higher education explores dozens of e-learning technologies with new ones seeming to emerge each week. Such technologies confront instructors and administrators at a time of continued budget retrenchments and rethinking. Adding to this dilemma, bored students are dropping out of online classes while pleading for richer and more engaging online learning experiences.

Given the demand for online learning, the plethora of online technologies to incorporate into teaching, the budgetary problems, and the opportunities for innovation, we argue that online learning environments are facing a "perfect e-storm," linking pedagogy, technology, and learner needs. Considering the extensive turbulence created by the perfect storm surrounding e-learning, it is not surprising that opinions are mixed about the benefits of online teaching and learning in higher education. As illustrated in numerous issues of the Chronicle of Higher Education during the past decade, excitement and enthusiasm for e-learning alternate with a pervasive sense of e-learning gloom, disappointment, bankruptcy and lawsuits, and myriad other contentions. Appropriately, the question arises as to where online learning is headed.

Navigating online education requires an understanding of the current state and the future direction of online teaching and learning. The study described here surveyed instructors and administrators in postsecondary institutions, mainly in the United States, to explore future trends of online education. In particular, the study makes predictions regarding the changing roles of online instructors, student expectations and needs related to online learning, pedagogical innovation, and projected technology use in online teaching and learning.

Review of Literature

We began this project with a review of past studies of the issues and trends in online teaching and learning in higher education.

Online Teaching and Learning

A recent survey of higher education in the United States reported that more than 2.35 million students enrolled in online courses in fall 2004. This report also noted that online education is becoming an important long-term strategy for many postsecondary institutions. Given the rapid growth of online education and its importance for postsecondary institutions, it is imperative that institutions of higher education provide quality online programmes.

The literature addresses student achievement and satisfaction as two means to assess the quality of online education. Studies focused on academic achievement have shown mixed reviews, but some researchers point out that online education can be at least as effective as traditional classroom instruction. Several research studies on student satisfaction in online courses or programmes reported both satisfied and dissatisfied students.

Faculty training and support is another critical component of quality online education. Many researchers posit that instructors play a different role from that of traditional classroom instructors when they teach online courses, as well as when they teach residential courses with Web enhancements. Such new roles for online instructors require training and support. Some case studies of faculty development programmes indicate that such programmes can have positive impacts on instructor transitions from teaching in a face-to-face to an online setting.

Pedagogy and Technology for online Education

Several research studies have covered effective pedagogical strategies for online teaching. Partlow and Gibbs, for instance, found from a Delphi study of experts in instructional technology and constructivism that online courses designed from constructivist principles should be relevant, interactive, project- based, and collaborative, while providing learners with some choice or control over their learning. Additionally, Keeton investigated effective online instructional practices based on a framework of effective teaching practices in face-to-face instruction in higher education.

In this study, Keeton interviewed faculty in postsecondary institutions, who rated the effectiveness of online instructional strategies. These instructors gave higher ratings to online instructional strategies that "create an environment that supports and encourages inquiry," "broaden the learner's experience of the subject matter," and "elicit active and critical reflection by learners on their growing experience base."

In another study of pedagogical practices, Bonk found that only 23–45 per cent of online instructors surveyed actually used online activities related to critical and creative thinking, handson performances, interactive labs, data analysis, and scientific simulations, although 40 per cent of the participants said those activities were highly important in online learning environments. In effect, a significant gap separated preferred and actual online instructional practices.

Technology has played and continues to play an important role in the development and expansion of online education. Accordingly, many universities have reported an increase in the use of online tools. Over the past decade, countless efforts have sought to integrate emerging Internet technologies into the teaching and learning process in higher education. Several studies have reported cases related to the use of blogs to promote student collaboration and reflection. Some researchers also have promoted the plausibility of using wikis for online student collaboration, and podcasting is beginning to garner attention from educators for its instructional use.

Although some discussions in the literature relate to effective practices in the use of emerging technologies for online education, empirical evidence to support or refute the effectiveness of such technologies, or, perhaps more importantly, guidance on how to use such tools effectively based on empirical evidence, is lacking.

Method

This study was based on a survey of individuals believed to have relevant experience with and insights into the factors affecting the present and future state of online education.

Participants

An online survey was conducted of college instructors and administrators who were members of either the Multimedia Educational Resource for Learning and Online Teaching (MERLOT) or the Western Cooperative for Educational Telecommunications (WCET), both premier associations for online education. MERLOT is a free and open resource for higher education with membership that, at the time of this study, included more than 12,000 college professors, instructional designers, and administrators who share and peer-evaluate their Web resources and materials (today, MERLOT has more than 35,000 members). WCET is an organization with 500-600 members that provides resources and information regarding the effective use of telecommu-nications technology in learning. Also surveyed were those who had posted one or more course syllabi at the World Lecture Hall (WLH), which has approximately 2,000 members and was developed by the University of Texas for faculty to share syllabi. This study is a part of a longitudinal effort to understand the use of technology in teaching, within both higher education and corporate training settings. The second author had previously surveyed MERLOT and WLH members on the state of online learning as well as corporate trainers on online training and blended learning.

Instrument

Using an online survey service, SurveyShare, we developed an online questionnaire as an instrument for this survey study. The questionnaire consisted of 42 questions grouped into three parts related to the current status and future trends of online education in higher education. The first part included 10 questions regarding respondents' demographic information. The second part included seven questions about the current status of online learning at the respondents' organizations. The third part included items regarding predictions about online teaching and learning. The survey used various types of questions, including Likerttype, multiple-choice, and open-ended questions.

Data Collection and Analysis

The survey took place from late November 2003 to early January 2004. An invitation was sent by e-mail to the sample of instructors, instructional designers, and administrators described earlier. The e-mail included information about the study as well as the URL to the survey site. Of more than 12,000 who received the e-mail request, 562 completed the survey. The participants responded to the survey anonymously, and the data were stored in the hosted online survey service. Descriptive data analyses (such as frequencies) were conducted using the data analysis tool provided in the online survey site.

Results

Our study confirmed some commonly held beliefs about online education, refuted others, and provided a range of predictions about the future of technology- enabled education.

Demographics of Online Instructors

Sixty-six per cent of the survey respondents held teaching positions (professors, instructors, or lecturers), while nearly one-fourth were administrators or instructional designers. Respondents represented institutions of various types: approximately half were employed by public, four-year colleges or universities; 23 per cent by community colleges or vocational institutes; and 16 per cent by private postsecondary institutions. A large majority (87 per cent) said their institutions offer online courses, and about 70 per cent of them had taught online courses. The experience with online teaching varied from none to more than 10 years.

Although not every respondent had online teaching experience, more than 95 per cent had experience integrating computer or Web technology into their face-to-face teaching. Survey results show that women appear to be teaching online in far greater numbers than just a few years ago. In fact, more than half of the respondents (53 per cent) were women. Such findings were surprising because a similar study conducted a few years earlier was dominated by male instructors who were full professors at tier-one universities. Perhaps female instructors had become more comfortable teaching and sharing activities online during the few years that elapsed between surveys, or perhaps support for instructors had improved on college campuses, or both.

Emerging Technology

When asked about several emerging technologies for online education, 27 per cent of respondents predicted that use of course management systems (CMSs) would increase most drastically in the next five years. Those surveyed also said that video streaming, online testing and exam tools, and learning object libraries would find significantly greater use on campus during this time. Between 5 and 10 per cent of respondents expected to see increases in asynchronous discussion tools, videoconferencing, synchronous presentation tools, and online testing.

The survey also asked what technology would most impact the delivery of online learning during the next five years. Respondents could select one of 14 key technologies. About 18 per cent of respondents predicted that reusable content objects and wireless technologies would have the most significant impact. Smaller percentages (from 7 to almost 14 per cent) selected peer-to-peer collaboration, digital libraries, simulations and games, assistive technologies, and digital portfolios.

In contrast, less than 5 per cent predicted that e-books, intelligent agents, Tablet PCs, virtual worlds, language support, and wearable technologies would have significant impact on the delivery of online learning. These findings seem to reflect the perceived importance of online technologies for sharing and using preexisting content. Additionally, respondents predicted that advances in Internet technology (for example, greatly extended bandwidth and wireless Internet connections) are likely to increase the use of multimedia and interactive simulations or games in online learning during the next five to 10 years.

Only about one in 10, however, predicted that advances in Internet technology would enhance videoconferencing or international collaboration, and just one in 16 thought it might offer greater chances to interact with field experts or practitioners. Again, the focus was on enhancing content and associated content delivery, not on the social interactions, cross-cultural exchanges, or new feedback channels that wider bandwidth could offer. Such responses indicate that respondents still see learning as content-driven, not based on social interactions and distributed intelligence. The emphasis remains on a knowledge-transmissio approach to education, not one rich in peer feedback, online mentoring, or cognitive apprenticeship.

Enormous Learner Demands

Our study revealed a number of trends related to areas of growth in online education, future needs for online instructors, and the dominance of online versus face-to-face instruction.

Growth of Online Programmes/Degrees

Comparing current online offerings and projected future online offerings at respondents' institutions yields predictions about the areas of growth in online programmes and degrees. Most respondents expected considerable growth in online certification and recertification programmes in the next few years, as well as in associate's degrees. Yet, our survey respondents predicted little growth in the number of institutions that offer online master's or doctoral programmes in the future. Although more than half of the respondents (54 per cent) expected that their institutions would offer online master's or doctoral programmes in the coming years, almost the same number of respondents (53 per cent) reported that their institutions were presently offering online master's or doctoral programmes. In contrast, respondents predicted that certification and recertification programmes would see 10–20 per cent growth from present offerings. Such responses indicate that higher education institutions might be wise to explore certificate and shortprogram offerings rather than full degree programmes.

Online Instructors' Readiness

Will online instructors be ready to meet the challenges brought by the projected increases in learner demands for online education? About half of the respondents predicted that monetary support for and pedagogical competency of online instructors would most significantly affect the success of their online programmes. In addition, instructors' technical competency was the third most pressing factor. Nevertheless pedagogical skill was deemed more important than technological skill for effective online teaching. With regard to the needs for pedagogical competency of online instructors, a majority of the respondents expected that online instructors would typically have received some sort of training in online teaching either internally or externally by the year 2010.

The Rise of Blended Learning

The survey asked respondents for their predictions related to the growth of online education in the next few years. Respondents indicated that more emphasis is expected on blended learning-instruction that combines face-to-face with online offerings-than on fully online courses. Those surveyed predicted a distinct shift from about onequarter of classes being blended today to perhaps the vast majority of courses having some Web component by the end of the decade.

Enhanced Pedagogy

Although the use of CMSs in higher education has increased rapidly and is likely the foundation for the rapid increase in the number of online learners during the past decade, some researchers argue that CMSs are promoted as ways to manage learners rather than to promote rich, interactive experiences. As a result, enhancing pedagogy is perhaps the most important factor in navigating the perfect e-storm. In the present study, respondents made predictions about the quality of online education in the near future and about how online courses would be taught and evaluated.

The Quality of Future Online Education

Survey respondents generally agreed with recent Sloan reports that the quality of online education will improve in the future. Sixty per cent of respondents expected that the quality of online courses would be identical to traditional instruction by the year 2006. Also, a majority of the respondents predicted that the quality of online courses would be superior to (47 per cent) or the same as (39 per cent) that of traditional instruction by 2013. Only 8 per cent predicted that the quality of online courses would be inferior in 2013. Similarly, a large majority of respondents predicted that learning outcomes of online students would be either the same as (39 per cent) or superior to (42 per cent) those of traditionally taught students by 2013.

In effect, the trend is for course quality and learner outcomes to steadily and significantly improve during the coming decade. Although we did not ask about reasons for the increase in quality, such numbers should be interesting and valuable to administrators, instructors, students, and other online learning stakeholders. In terms of factors that can improve online learners' success, respondents said that training students to selfregulatetheir learning (22 per cent) was needed most, followed by better measures of student readiness (17 per cent), better evaluation of student achievement (17 per cent), and better CMSs to track student learning. Nine per cent said additional technology training is needed. This concern about learner self-regulationis ironic in a world dominated and driven by learning management systems that are primarily used to manage students, as alluded to earlier. Follow-up surveys might address whether learners per- ceive this mixed message and whether they prefer to be managed online or engage in more self-directed online environments.

As Carmean and Haefner argued, there is a need for CMS environments that foster deeper student learning and engagement. They noted that such environments might foster student choice among various activities, reflection, apprenticeship, synthesis, real-world problem solving, and rich, timely feedback. More recently, Weigel added to this argument by suggesting that the next-generation CMS should foster a more learner-centered environment that rich in critical thinking, student exploration, peer learning and knowledge construction, interdisciplinary experiences incorporating a community of educators (practitioners, business leaders, alumni, and others), and educational opportunities.

Online Teaching Skills

Instructors' abilities to teach online are critical to the quality of online education. Unlike our earlier study related to the state of online learning in 2001, which included many questions about online learning tools and features, the present study focused more on learning outcomes and pedagogical skills. For instance, this study found that the most important skills for an online instructor during the next few years will be how to moderate or facilitate learning and how to develop or plan for high-quality online courses. Being a subject-matter expert was the next most important skill. In effect, the results indicate that planning and moderating skills are perhaps more important than actual "teaching" or lecturing skills in online courses. As Salmon pointed out, online instructors are moderators or facilitators of student learning.

Pedagogical Techniques

Over half of the survey respondents predicted that online collaboration, casebased learning, and problem-based learning (PBL) would be the preferred instructional methods for online instructors in the coming decade. In contrast, few respondents expected that instructors would rely on lectures, modeling, or Socratic instruction for their online teaching in the future. In other words, survey respondents predicted that more learner-centered techniques would be used in the future, indicating a marked shift from traditional teacher-directed approaches. Existing research indicates that online instructors tend to use easy-toimplement tools, resources, and strategies rather than complex PBL, virtual teaming, cross-cultural collaboration, simulations, and other forms of rich interactive media.

If the prediction for more learner-centered pedagogies online is realised, it would be interesting to study whether those teaching online transfer such pedagogical skills to their face-to-face instructional activities. Our findings also indicated that, in general, respondents envisioned the Web in the next few years more as a tool for virtual teaming or collaboration, critical thinking, and enhanced student engagement than as an opportunity for student idea generation and expression of creativity. This is not surprising, given that most instruction in higher education is focused on consumption and evaluation of knowledge, not on the generation of it. Perhaps online training departments and units need to offer more examples of how to successfully embed creative and generative online tasks and activities.

Evaluation and Assessment of Online Courses

Evaluation is an important part of ensuring the quality of online courses and programmes. When asked how the quality of online education will be most effectively measured during the coming decade, 44 per cent answered that a comparison of online student achievement with that of students in face-to-face classroom settings would be the most effective, followed by student performance in simulated tasks of real-world activities (15 per cent), calculations of return on investment (10 per cent), and student course evaluations (9 per cent).

Clearly, respondents believe that face-to-face instruction provides a valid benchmark for teaching and learning outcomes and that online performance should at least equal its effectiveness. Such views, while politically important, seem to forget that much of the learning that occurs online could not take place in a faceto- face delivery mode (for example, asynchronous online discussions or online mentoring). It also assumes that face-to-face instruction is superior. What if institutions took the opposite stance and measured face-to-face courses based on whether they couldaccomplish all that online instruction can?

As for the forms of evaluation that will be used during the next few years, respondents predicted that online practice quizzes and exams would be most highly used, followed by online surveying and polling, course evaluations, and online quizzes and exams. In particular, more than 90 per cent of the respondents predicted that online surveys would be used as an important student research tool or as a teaching device in addition to student assessment and course evaluation. This finding affirms our belief that online surveys offer the chance to be learner-centered because they allow students to collect, analyse, and report on real-world data and projects.

Chapter 7

Human Resources Management Information Systems

In the increasingly high-tech world we live in, more HRM activities will require specialised expertise. For example, many organisations are developing computerized expert systems for making employee selection decisions. These systems integrate interview data, test scores, and application form information. Similarly, many organisations are developing compensation systems with elaborate cafeteria-style benefit packages to replace simple hourly pay or piece-rate incentive systems. Many organisations are developing sophisticated databases to centralize all HRM information to provide real-time information for strategic human resources planning and other reporting activities.

In concluding this chapter it is important to remind the reader that regardless of their size, mission, market, or environment, all organisations strive to achieve their goals by combining various resources into goods and services that will be of value to their customers. Financial resources such as ownership investment, sales revenues, and bank loans are used to provide capital and to cover expenses necessary to conduct business. Material resources such as factories, equipment, raw materials, computers, and offices play an important role in the actual creation of goods and services.

And information resources about consumers and the organisation's competitive environment help managers make decisions, solve problems, and develop strategies. But no resources are more vital to an organisation's success in today's world of work than human resources. Organisations that once paid lip service to human resources issues are increasingly recognising the dramatic impact that effective HRM can have in all areas of an organisation. Indeed, effective HRM has become a vital strategic concern for most organisations today, and there is every indication that the trend will continue in the years to come. Effective management of human resources is an important component of organisational success. And the HRM function is an organisation's most critical source of information about employment practices, employee behaviour, labor relations, and the effective management of all aspects of human resources. Human resources management is made up of an identifiable set of activities that affect and influence the people who work in an organisation. These activities include HRM planning, job analysis, recruitment, selection, placement, career management, training and development, designing performance appraisal and compensation systems, and labor relations.

The current challenge of HRM is to integrate programs involving human resources with strategic organisational objectives. More and more, organisations are under tremendous competitive pressure worldwide. HRM personnel must find ways to develop effective international programs that meet this challenge. The challenges for HRM today and tomorrow involve responding to corporate reorganisations, global competition, increasing diversity in the workforce, employee expectations, and legal and governmental requirements in order to help their host organisations meet their organisational goals. In the end, HRM must help make organisations better, faster, and more competitive.

Strategic Role of Information Systems

Over the past decade, the resource-based view of competitive advantage has emerged as a popular approach to examining the strategic roles of information systems (IS). One critical issue in the resource-based inquiry of the strategic impacts of IS is whether IS alone can lead to competitive advantage or they must work in conjunction with other organisational resources in order to provide strategic benefits. The former suggests a direct effect of IS on firm performance, whereas the latter implies an indirect effect of IS. While researchers have increasingly embraced the latter by arguing that IS complemented by certain organisational resources may lead to competitive advantage and superior performance, there has been relatively less empirical attention to testing the indirect effect of IS. Since the indirect effect of IS has become more and more influential in current thinking of how to evaluate and manage IS resources, more empirical evidence is needed to ascertain this effect.

Furthermore, even though the indirect effect of IS generally exists, we still don't know enough about what specific organisational resources complement IS in influencing a firm's competitive position or performance. While the normative literature proposes a number of potential organisational complements to IS, the performance impacts of many of those complementary resources have not been assessed in prior empirical research. Discerning the influence of different IS complements on the IS-performance relationship would then increase our knowledge of what represents a relevant set of complementary resources that interact with IS in affecting firm performance.

The purpose of this study was twofold. First, it provided another assessment of the indirect effect of IS on firm performance. Second, by testing the relationships between three sets of firm-specific complements to IS and firm performance, the study sought to empirically determine what organisational resources complement IS in influencing firm performance. The remainder of the chapter is organised as follows. The next section reviews the indirect effect of IS within the resource-based research on IS impacts, as well as the existing empirical evidence. This is followed by an examination of the potential performance impacts of three types of organisational resources (unique organisational culture and structure, unique vertical integration and related diversification, and unique knowledge and information) that complement IS. The methodology section describes the empirical analysis, including the sample and data collection procedure, the measurement of the variables of interest, and the results. The discussion section presents the implications of the research findings, the limitations of the study, and some suggestions for future research and practice. The last section provides a summary and conclusions for the study.

As a popular theoretical perspective in the strategic management literature, the resource-based view of competitive advantage suggests that firms with unique and difficult to imitate or substitute resources and capabilities can gain sustainable competitive advantage and superior performance. Over the past decade, the resource-based research has placed increasing emphasis on bundling a firm's resources and capabilities in creating and maintaining competitive advantage, for example, have developed and used the notion of payments (costs to a resource) to show that superior organisational performance is achieved by finding the most valuable combination of a firm's resources and bargaining over the marginal contribution of combining the resources.

Drawing upon the concept of resource complementarity (the presence of a resource enhances the strategic values of other resources it complements), resource-based researchers further posit that firms exploiting the complementarity among their resources and capabilities can create complex resource /capability networks as barriers to imitation, thus enhancing the potential of achieving durable competitive advantage. Recent empirical studies have shown that the combinative effects of complementary resources and capabilities influence the competitive performance of firms. Song et al., for instance, found a synergistic effect between two complementary organisational capabilities (marketing-related and technology-related) on firm performance in the high turbulence environment.

The Resource-based View of the Strategic Roles of IS

Since the early 90s, IS researchers have turned to the resource-based view in examining the strategic roles of IS and explaining the 'productivity paradox" regarding the strategic impacts of IS. The early resource-based analyses viewed IS as commodity-like resources that are generally neither unique nor difficult to imitate, hence rarely resulting in sustainable competitive advantage. In the literature, this perspective is known as the "strategic necessity hypothesis".

While acknowledging that the direct effect of IS rarely exists, more recent resource-based inquiry has shown that IS may still have an indirect effect on a firm's competitive position or performance. That is, despite lacking the characteristics required for sustainable competitive advantage, IS may exert positive influence on firm performance through their relationships with other organisational resources. Following the logic of resource complementarity, IS and strategy researchers have argued that firms whose IS are complemented by other firm-specific and hard-to-copy organisational resources are in a better position to defend their IS-derived competitive advantage than those that lack such resources. According to this line of reasoning, though the necessary software and hardware used by a firm's IS can be easily imitated, it is more difficult for its competitors to copy the unique and intangible resources the firm uses in implementing and exploiting its IS. Moreover, blending IS with other organisational resources may create a complex set of complementary resources that are not easily matched by competitors, thus sustaining IS-based advantage.

Despite gaining acceptance among researchers who analyze the strategic impacts of IS from the resource-based perspective, the indirect effect of IS has not been subject to close empirical scrutiny. Since it was first proposed in the early 90s, the indirect effect of IS has been tested in only two studies. In a longitudinal study of thirty IS considered as 'classic' cases of strategic use of information technology in the literature, Kettinger et al. explored the potential influence of a number of organisational factors on the sustainability of the IS-based competitive advantage. They found an established technological base and substantial capital availability as two main organisational resources that differentiated the IS producing sustained superior performance from the IS resulting in only temporary superior performance. These findings seem to provide initial evidence for the indirect effect of IS, although the effects of several resources (e.g., competitive scopes and information resources) that might potentially affect IS-derived advantage were not tested due to lack of data availability.

In a subsequent study, Powell and Dent-Micallef investigated the indirect effect of IS with cross-sectional data collected from 67 U.S. retailers. They examined and tested the relationships between three sets of organisational resources (information technology resources, complementary human resources, and complementary business resources) and perceived firm performance. The complementary human resources under study included open organisation, open communications, organisational consensus, CEO commitment, organisational flexibility, and IT-strategy integration. The complementary business resources encompassed supplier relationships, IT training, business process design, team orientation, benchmarking, and IT planning. Powell and Dent-Micallef found that IT resources alone did not explain significant firm performance. They further found that some retailers gained performance advantages from using complementary human resources. Since the empirical testing of the study focused on bundles of complementary resources, the performance effects of individual complementary resources were not closely checked. Moreover, it remains unclear whether the complementary human resources under study were unique to the firms.

While generally supporting the indirect effect of IS, the empirical works by Kettinger et al. and Powell and Dent-Micallef were insufficient in identifying a relevant set of complementary organisational resources contributing to the indirect effect of IS. The following sections examine three types of distinctive organisational resources (unique organisational culture and structure, unique vertical integration and related diversification, and unique knowledge and information) and their interactions with IS in affecting firm performance. These complementary resources were selected as the loci of this study because they not only may contribute to IS-based competitive advantage, but also tend to be firm-specific and hard to copy. As noted above, a firm is in a better position to protect its IS-based advantage if its IS are complemented by other organisational resources that are idiosyncratic to the firm and difficult to imitate.

Unique Organisational Culture and Structure that Complement IS

Organisation and strategy researchers have long recognised the key roles of organisational culture and structure in developing and leveraging resources and capabilities for competitive advantage. Barney argues that, without supportive organisational culture and structure, a firm is less likely to exploit the full competitive potential of its resources and capabilities. Miller showed how Citicorp, under the leadership of John Reed, used different cultural and structure mechanisms (e.g., a collaborative culture, project teams and cross-functional committees) to turn its international branch network into a source of sustainable competitive advantage. Besides complementing other organisational resources and capabilities, organisational culture and structure tend to be imperfectly imitable. It is well recognised in the resource-based literature that idiosyncratic and valuable organisational cultures are difficult to duplicate because they represent socially complex phenomena.

As Barney notes, even though firms lacking certain attributes of a valuable organisational culture may understand how these attributes contribute to competitive performance, systematic efforts to create those attributes typically require simultaneous manipulation of complex social relationships, hence making imitation costly. Research into the organisational impacts of organisational structure has also shown that duplicating effective organisational structures is difficult in that they are often context-bound (i.e., they must be properly matched with the particular organisational situations) and require synergistic integration of different organisational elements (e.g., processes, systems and capabilities). To illustrate the complexity involved in designing and adopting an organisational structure, Galbraith makes the following observation regarding the difficulty multinational firms may face in designing a transnational structure:

"It is not simply a matter of distributing regional and global mandates. Rather it involves the creation of global management teams for the top group, as well as other key groups like product-development teams; the design of global business processes and information systems; the creation of new measurement and reward systems; and, finally, the use of managers with a global mind-set and team skills."

It is evident from several streams of research that a firm's organisational culture and structure are instrumental in influencing its ability to derive strategic benefits from IS. The absence of organisational culture and structure supporting the smooth implementation and use of IS has been documented as a major cause of many system failures in the IS implementation and adoption literature. Several empirical studies, for instance, have found relatively low system use among firms lacking a culture and reward systems that support IS adoption. The business process reengineering research also demonstrates that firms whose structures and processes are not aligned with their new IS have experienced difficulty in reaping the benefits of the IS. Moreover, recent research on organisational barriers to knowledge management suggests that firms may not be able to turn data and information into useful knowledge and organisational results from their IS without a supportive organisational culture and structure. Even if new knowledge is created from employing IS, sharing the new knowledge may be limited by cultural and structural constrictions. Aside from affecting the economic impacts of IS, firm-specific organisational culture and structure make it difficult for competitors to imitate the IS they complement because organisational culture and structure tend to be socially complex and hence difficult to imitate, as noted above.

Hypothesis 1: IS complemented by unique organisational culture and structure are positively related to firm performance.

Unique Vertical Integration and Related Diversification That Complement IS

Strategy scholars have long argued that competitive advantage can be achieved with competitive scope. Among several dimensions of competitive scope are vertical integration and related diversification. The former describes the extent of a firm's integration into the businesses of its buyers or suppliers, while the latter refers to the range of related businesses the firm competes in. Both vertical integration and related diversification have been shown as potential sources of sustainable competitive advantage in the strategic management literature. Research based on transaction cost economics, resource-based theory and knowledge-based view of the firm suggest that vertical integration may confer economic value by allowing the firm to avoid market exchange costs arising from opportunism, uncertainty and asset specificity and better manage and utilise its unique and hard-to-copy skills and knowledge in certain functions. Moreover, Barney has recently examined the roles of governance skills in vertical integration and argued that firms with superior governance skills (e.g., ability to analyze uncertain and complex economic transactions) that are rare and costly to imitate may increase and sustain competitive advantage derived from vertical integration.

The related diversification literature indicates that related diversification based on sharing related activities or competencies is generally associated with superior firm performance. Further, certain types of related diversification tend to be more firm-specific and harder to duplicate. In his analysis of the potential linkage between related diversification and sustainable competitive advantage, Barney notes that related diversification that exploits rare and costly to imitate economies of scope (e.g., core competencies) is more unique and immune from direct imitation than one based on common and less costly to imitate economies of scope (e.g., shared activities and risk reduction).

IS researchers adopting the resource-based perspective note that firms can create and maintain competitive advantage from merging IS with unique related diversification or vertical integration. Clemons and Row identify three ways IS can be deployed to exploit a firm's unique diversification scope for competitive advantage. First, firms with wider ranges of related businesses may achieve a scale advantage from using IS to improve coordination of similar activities and resources across various markets, putting their rivals with more limited business scopes at a cost disadvantage. Second, by facilitating the transfer and sharing of critical skills and knowledge among multiple businesses, IS give a firm with a broader industry scope better leverage of its expertise and hence competitive advantage. Third, IS may be used to generate synergistic effects (i.e., creating more value for the customer from combining different, but complementary resources in different lines of business).

Firms may also exploit variations in vertical integration to derive more performance benefits from IS. A firm performing more vertically related activities can design and deploy IS to leverage unique information and knowledge resources from its upstream or downstream businesses, hence creating an advantageous position over its less vertically integrated competitors. In a classic case, Otis (once an independent company of elevator manufacturing and service) installed a remote diagnostic computer system in the elevators it produced to capture and provide critical information to its service database. Such unique information created by the system enabled Otis to obtain competitive advantage over other elevator service providers.

On the other hand, firms with shorter value chains may use IS to form a network of "quasi-vertical" or "virtual" integration with their trading partners. IS-based virtual integration allows individual firms within the network to enjoy operational benefits of vertical integration (e.g., higher efficiency and increased coordination), while also reducing the transaction costs and risks associated with vertical integration and realising the production economies available to separate, specialised firms. Although competitors with full vertical integration may potentially match the level of operational integration, it is not as easy for them to match the production economies and flexibility of independent and specialised firms that are connected together by IS.

Hypothesis 2: IS complemented by unique vertical integration and related diversification are positively related to firm performance.

Unique Knowledge and Information That Complement IS

It is widely recognised today that knowledge and information represent the most important resources of competitive advantage. Knowledge and information not only increasingly add value to products and services, but also play a vital role in transforming resources and capabilities into dynamic core competencies. Moreover, because organisational knowledge tends to be tacit, socially complex, embedded in firm-specific routines and processes, and nontradeable in strategic factor markets, the knowledge-based advantage is difficult to copy and thus sustainable. Like knowledge, firm-specific information can be hard to imitate. For instance, proprietary databases (e.g., customer databases) may take years to build, and their development and access are often specific to a firm's interactions with its business environments.

It is evident in the literature that the successful implementation and exploitation of IS to achieve such operational benefits as production efficiency, product flexibility and close cross-functional coordination depend upon a firm's knowledge resources. Research on advanced manufacturing technologies (AMT) such as computer-aided design (CAD) and computer aided manufacturing (CAM) demonstrates that the richness of a firm's tacit knowledge (the insights, heuristics and experience of the firm's employees) applied in the procedures and workflows supported by AMT influences the long-term implementation success of AMT.

For example, Upton (1995) argues that manufacturers with workers adept at carrying out quick changeovers and responding to the demands of new customers are more likely to create a manufacturing system that combines IT and employee skills to make IS-based flexibility work. Parthasarthy and Sethi posit that firms whose employees possess the skills for selecting, processing and transmitting complex information quickly would enjoy greater economic gains from

IS-based flexibility. Kotha's (1995) case study of how National Bicycle Industrial Company (NBIC), a Japanese bicycle manufacturer, developed and implemented mass customization for competitive advantage revealed that access to highly trained workers and substantial in-house expertise in engineering and manufacturing played a critical role in NBIC's ability to develop and deploy IS to offer a great variety of bicycles at low costs. The study also showed that the same knowledge resources enabled NBIC to use IS to integrate different functional activities and establish a close information network with its customers and suppliers.

The competitive advantage derived from combining human expertise with IS is harder to duplicate because employee skills and knowledge that complement IS are often unique and contingent on firm-specific organisational routines developed over an extended period of time. In their resource-based analysis of several IT-related resources, Mata et al. concluded that managerial skills in building, implementing and managing IT are rare among firms, require long periods of practice and learning, and involve complex social relations. The mass customization experience of NBIC mentioned above showed that the main rivals of NBIC had a hard time trying to imitate its approach to mass customization because NBIC's IS that supported its mass customization operation was built with in-house engineering and manufacturing expertise accumulated over many years. Furthermore, firms blending their IS with unique knowledge resources may be able to create a complex set of complementary resources that are not easily matched by competitors.

A firm is also in a better position to derive IS-based advantage if the firm possesses unique information resources. The presence of a proprietary database may create more strategic opportunities the firm can exploit with its IS. For instance, Kraft General Foods developed a repertoire of usable promotion programs, products, value-added ideas, and selling tools from employing a centralized IS to access and analyze sales and consumer data collected from 30,000 food stores nationwide. In a more recent example, Boston's Fairmont Copley Plaza Hotel provides its concierges with a guest-history database and street information through a computerized system to expedite guest service at the concierge desk. This IS support has consistently boosted the concierge's guest-satisfaction index close to 90% and promoted loyalty among the hotel's core group of guests. Further, the possession of firm-specific information not only increases the value of IS, but also makes imitation difficult. While competitors may build similar IS easily, it is harder for them to develop the comparable database that may take a long time to build.

Hypothesis 3: IS complemented by unique knowledge and information are positively related to firm performance.

Sample and Data Collection

The data for this study came from two sources. The data tapping the independent variables were gathered via a mail survey administered in 1998, and the data about the performance and control variables were obtained from the Research Insight (formerly known as Compustat) database. The target respondents of the survey were senior IS executives in large (Fortune and Forbes) firms in the U.S. Most of the respondents held the positions of either vice presidents of IS or chief information officers. The senior IS executive was chosen as the single informant in this study because of his or her familiarity with both IS and strategic management issues. Several previous studies have found increasing involvement of senior IS executives in strategic planning and control activities of firms. Furthermore, a recent study found the information offered by key IS executives consistent with the insights obtained from other senior members of management. Hence, IS researchers have increasingly relied on senior IS executives as single informants in gathering data about strategic IS issues.

The contact information of the senior IS executives was obtained from the Directory of Top Computer Executives compiled by Applied Computer Research Inc. From this source, a sample of 879 firms that had financial data in the Research Insight database was identified. Before being mailed to the target respondents, the survey instrument was pre-tested and refined for content validity and item clarity with senior IS executives from five Fortune 500 companies headquartered in a mid-western state. One hundred and one questionnaires were undelivered or returned because the IS executives were no longer with the companies. Twenty-nine firms declined to participate in the study in writing, on the phone, or through e-mail. To boost the response rate, two follow-up mailings and one reminder letter were initiated after the first mailing. Of the 778 firms that received the questionnaires, a total of 164 responses were received, out of which 16 responses were unusable. The effective response rate was thus 19% (148 responses). Such a response rate is comparable to those reported in similar studies using senior IS executives in large firms.

To test for potential nonresponse bias, the respondent firms were compared to their non-respondent counterparts with respect to sales and number of employees. T-test results showed no significant differences between the two groups: sales and number of employees. In keeping with Armstrong and Overton, another nonresponse bias check was conducted by comparing early with late respondents. T-tests of the mean differences for the three explanatory variables failed to reveal any significant differences: unique organisational culture and structure that complement IS, unique vertical integration and related diversification that complement IS, and unique knowledge and information that complement IS. Together, these checks provided some evidence for the absence of non-response bias in the data set.

Measures

Independent Variables

Based on the works by Feeny and Ives, Clemons and Row, and Kettinger et al., six items were developed to measure the six different unique organisational resources that complement IS. For each of the six items, the respondents were asked to indicate the extent to which the use and implementation of their IS required each of these resources on a five-point, Likert-type scale with anchors ranging from "Very great extent" (= 5) to "No extent" (= 1). To assess the construct validity and unidimensionality of the scale, a principal components factor analysis with varimax rotation was performed on the six items. The factor analysis revealed three factors explaining about 77.8% of the total variance and corresponding with the three proposed sets of complementary resources, respectively.

Dependent Variables

Two popular measures of profitability, return on sales (ROS) and return on assets (ROA), were employed to measure the performance impacts of the unique organisational resources that complement IS. Both profitability ratios have been frequently used in previous assessments of the strategic impacts of IS. To smooth annual fluctuations and reduce short-term effects to some degree, a two-year average was used for both variables.

Control variables. Since the firms participating in this study came from a variety of industries, it was necessary to control, to some degree, the different industry conditions under which the firms operated. To control for the industry effects, SIC codes were first used to classify the firms into four groups: 1) manufacturing, 2) transportation and public utilities, 3) wholesale and retail trade, and 4) service. Where a firm operated in more than one industry, the firm's SIC code was determined by identifying the industry from which the firm received the largest percentage of sales and the corresponding SIC code. Three dummy variables (each with values of 0 or 1) were then created for the second (transportation and public utilities), the third (wholesale and retail trade) and the fourth (service) groups of firms. For each dummy variable, a firm was assigned a value of 1 if it belonged to a group.

The fourth control variable was firm size, which has frequently been used in previous studies involving firm performance as a dependent variable. In keeping with Kettinger et al., firm size was measured with total assets. The fifth control variable was technological resources. A firm's technological resources may influence its ability to develop IS for sustainable competitive advantage. While a preferable measure of technological resources is R&D intensity, the Research Insight data for R&D intensity were missing for many firms in the sample. An alternative measure (investment intensity operationalized as invested capital to sales), as recommended by Kettinger et al., was then used for technological resources. The last two control variables represent two types of financial slack: available slack and potential slack. Reflecting a firm's ability to generate cash flow for reinvestment, financial slack needs to be controlled due to its influence on the firm's financial performance as well as its ability to invest in and develop IS. Following convention, available slack was measured as the current ratio (current assets to current liabilities) and potential slack as the debt to equity ratio.

Analyses

To test the hypotheses, two sets of two-stage regression analyses were performed, using ROS and ROA as the dependent variables. In the first stage of each set of the analyses, the seven control variables were entered into the regression model as a set. In the second stage, the three independent variables were added to the equation. To avoid potential multicollinearity among the three independent variables, their factor scores were calculated from the factor analysis and used in the regression analyses.

Unique knowledge and information that complemented IS were positively correlated with ROS, while unique vertical integration and related diversification that complemented IS were positively associated with ROA. Unique organisational culture and structure that complemented IS were not significantly correlated with either ROS or ROA. Hypothesis 1 predicts that IS complemented by unique organisational culture and structure are positively related to firm performance. Therefore, Hypothesis I was not supported. Hypothesis 2 states that IS complemented by unique vertical integration and related diversification are positively related to firm performance. The same models reveal that unique vertical integration and related diversification that complemented IS were positively related to ROA ($b = .18$, $p < .05$), but not ROS. Hypothesis 2 was thus partially supported. Hypothesis 3 suggests that IS complemented by unique knowledge and information are positively related to firm performance. As shown in Models 2 and 4, unique knowledge and information that complemented IS were positively associated with both ROS ($b = .21$, $p < .01$) and ROA ($b = .19$, $p < .05$), hence supporting Hypothesis 3.

Discussion

This research sought to empirically test the indirect effect of IS on firm performance and identify several firm-specific, complementary organisational resources contributing to that effect. The results indicate that firms whose IS were complemented by unique knowledge and information enjoyed gains in ROS and ROA. IS complemented by unique vertical integration and related diversification could also lead to higher ROA. Consistent with the normative literature and the empirical work by Kettinger et al. and Powell and Dent-Micaleff, these findings provide additional evidence in support of the resource-based argument that IS influence on firm performance arises from their interactions with other firm-specific and hard-to-copy organisational resources. While confirming the indirect effect of IS, this study differed from the previous studies by finding empirical support for the roles of unique knowledge, information, vertical integration and related diversification that complemented IS in affecting the relationship between IS and firm performance. One possible fruitful extension to the research on the indirect effect of IS is to identify and examine other distinct organisational resources that could potentially enhance IS impacts on firm performance.

Another possible direction for future inquiry in this line of research is to investigate the interrelationship between IS and organisational knowledge/information. Since IS are capable of helping firms develop valuable and firm-specific organisational knowledge and information that in turn can be used to facilitate the implementation and utilisation of IS, a firm can create a reciprocal relationship between these two types of organisational resources, which could then increase the complexity of the resource complementarity and hence make imitation more difficult.

Contrary to the expectations and the findings by Powell and Dent-Micaleff, the study found no evidence for the performance influence of firm-specific organisational culture and structure that complemented IS. This unexpected non-finding might be due to the coarse measures used in the study. Another possible explanation is that some unique organisational cultures and structures in the sample might not exhibit certain desirable characteristics such as open organisation and open communications. Consequently, even if those organisational cultures and structures were perceived as firm-specific and complementary to IS, their interactions with IS did not exert positive influence on firm performance. It then appears that more in-depth studies that draw from comprehensive analyses of organisational culture and structure are needed to identify specific aspects or types of organisational culture and structure, which are not only conducive to the implementation and exploitation of IS, but also unavailable to competition.

Managerial Implications

The findings from this research have practical implications for the strategic management of IS. While firms these days are investing heavily in building and deploying IS to improve their competitive positions, the performance impacts of such IS investments depend on the presence of certain firm-specific resources that complement the IS. A firm is more likely to reap economic benefits (gains in profitability) from its IS investment if it possesses firm-specific knowledge, information, vertical integration and related diversification that facilitate IS implementation and exploitation. Hence, creating and utilising unique knowledge and information that increase the effectiveness of IS investments are as important as making the IS investments. Moreover, aligning IS with a firm's unique vertical integration and related diversification may increase the performance contributions of the IS.

The results presented here can be interpreted to imply a larger role for IS in helping firms gain competitive advantage than that suggested by those who question the strategic value of IS. Contrary to the growing skepticism towards whether IS can be more than a "strategic necessity," the findings suggest that IS can be a source of competitive advantage and superior economic performance if they are complemented by certain distinct organisational resources. Accordingly, the critical issue facing firms and their managers is not whether they should invest in IS, but how to manage the complementarity between IS and other organisational resources to maximize IS payoffs.

The findings in this study need to be interpreted within its limitations. First, the study relied on perceptual data collected from single informants in measuring organisational resources that complemented IS. Data collected in such a manner might be influenced by the respondents' cognitive biases and distortions, although objective measures were used to reduce similar biases and inaccuracies in collecting the data for the performance and control variables and avoid potential common method variance. Another measurement limitation lies in the use of single-item scales to measure the IS complements. These general measures might be insufficient to fully capture the complexity in certain complementary resources (e.g., unique organisational culture, and unique organisational structure) and subject to different interpretations by different respondents. The coarseness of the measures might then have contributed to the non-finding for unique, complementary organisational culture and structure as well as the low reliability of unique, complementary knowledge and information. Therefore, future research on the performance impacts of complementary organisational resources of IS need to develop and use multi-item scales with higher validity and reliability to measure these resources.

While the study controlled for a number of industry and organisational factors, there might be other potential performance determinants whose effects were not taken into account due to the lack of data and the small sample size. The exclusion of those variables might have resulted in overestimating or underestimating the contributions of the unique, complementary organisational resources to the indirect effect of IS. Whenever possible, future studies need to include other environmental and organisational attributes related to firm performance in order to obtain more accurate assessments of how unique, complementary organisational resources interact with IS in affecting firm performance. As another limitation, the response rate (19%) for the survey used in this research was relatively low. While comparable to those of similar studies, this response rate may limit the generalizability of the study results. Obtaining a high response rate for sensitive information concerning the strategic use of IS continues to be a challenge for researchers.

This study tested the indirect effect of IS on firm performance, which has received increasing attention in the resource-based research on the strategic roles of IS. Among three potential types of complementary organisational resources contributing to the indirect effect of IS, the study found that unique knowledge and information complemented IS in improving profitability. Although to a lesser degree, the study also found the presence of unique, complementary vertical integration and related diversification positively associated with profitability. On the other hand, unique organisational culture and structure which have been often deemed as critical to IS effectiveness and contributions to firm performance were not found to have any significant effect. Together, the results from this study not only provide empirical evidence that the indirect effect of IS may exist, but also increase our knowledge of IS complements that are more likely to contribute to IS-based competitive advantage. While representing one of the few empirical endeavours to assess the indirect effect of IS and identify what types of complementary organisational resources contribute to that effect, this study suggests that additional research based on more rigorous methodology is needed to help us fully understand what represent a relevant set of IS complements that affect the IS-performance relationship.

Information as a Competitive Weapon

The need to integrate technology into the strategic plans of a firm is widely recognised as vital to the health and longevity of the company. The firm's technologies and the research for advancing them will be more effective if they are linked to market needs and customer demands. The major factor that prevents many firms from achieving their technical objectives and, therefore, their strategic objectives, is the lack of resources. For technology research and development (R&D), the insufficient resources are usually capital and technical "critical mass." The costs of building and sustaining the necessary technical expertise and specialised equipment are rising dramatically. Even for the largest corporations, such as Fortune 100 firms, leadership in some market segments they have traditionally dominated cannot be maintained because they lack sufficient technical capabilities to adapt to fast-paced market dynamics.

In response, technology partnerships between and in some cases among organisations are becoming more important and prevalent. From 1976 to 1987, the annual number of new joint ventures rose six-fold; by 1987, three-quarters of these were in high-technology industries. As the costs, including risk associated with R&D efforts, continue to increase, no company can remain a "technology island" and stay competitive. Today's strategic planners must broaden their view of their business environment from the traditional perspective of individual firms competing against each other. The formation of strategic alliances means that strategic power often resides in sets of firms acting together. Because competitors are capitalizing on the benefits of partnerships, they must be considered an option in the planners' array of alternatives.

Strategic Alliances

A strategic alliance, broadly defined, is a contractual agreement among firms to cooperate to obtain an objective without regard to the legal or organisational form the alliance takes. This definition accommodates the myriad arrangements that can range from handshake agreements to licensing, mergers, and equity joint ventures. Thus, strategic alliances extend to all relationships within the marketplace. Firms can strengthen current relationships with customers, suppliers, distributors, academia, and even competitors, as each plays a major role in whether a firm meets its strategic objectives. Alliances to achieve common objectives accentuate intercompany dependencies. A growing interdependence among key strategic partners is vital to continued participation and prosperity in the global economy.

Alliances permit smaller firms to leave the niche markets to which they have been relegated and face the dominant market players head-on. By aligning themselves with firms that possess a resource needed for expansion, small firms can capitalize on their own strengths to a much greater extent. Stiff entry barriers to new markets can be overcome as alliances gain access through complementary internal assets. Generally, firms must first define what "core" technologies give them the greatest competitive advantages. Then, to achieve the maximum leverage of these strengths, they must seek peripheral capabilities from the external environment. In today's environment, few firms can afford the luxury, and possibly the resources, to build or develop adequate internal technologies or capabilities necessary to seize all new market opportunities. This is especially true in a global marketplace where response times are extremely short.

Different guidelines and reasons for partnership development with various strategic partners must be considered by management. A later section will focus on selected business relationships and why a firm would consider a specific alliance. The success or failure of a strategic alliance lies directly with the management competence in the allied firms. Alliances involve shared risks grounded in mutual objectives and trust. The traditional organisational hierarchy controlled by typical management philosophies will not be effective in realising the power inherent in collaborative efforts. Strong working relationships and total commitment with downward delegation of decisions to those ultimately responsible are hallmarks of the successful alliance manager. In essence, alliances require a redefinition of the firm and its corporate culture by the very people who in the past have found the old definition quite profitable.

Considerations for Forming Alliances

The decision to form a technology alliance is viable under many circumstances. Much depends upon how the alliance is forged and managed. While the decision to enter into an alliance may be for one of the reasons outlined, there are still pitfalls for the unwary manager. These warnings are based on observations of collaborative projects that failed. Major technological breakthroughs often require resources to develop and commercialize that are beyond the scope of a single firm. If the necessary human, financial, and physical capital exceeds the resource base of the company, a partner can be solicited.

Both gain from the reduced individual development effort and new access to existing technology. Creating a critical mass for R&D can be achieved more quickly by enlisting scarce resources from other firms already experienced in some aspects of the development and application of the technology. The firms can now proceed more deliberately to develop a higher technical base through experiential learning while still satisfying short-term demands of business opportunities. Furthermore, entrance into new businesses or markets is facilitated through the alliance of firms with complementary financial, scientific, and cultural strengths.

Sharing risk is another important reason to enter a technology alliance. The risk of R&D failure has always been a concern, and sharing the risk eases the swain. Ironically, a second and potentially greater risk is associated with successful commercialization. The risk lies in having an unknown social or environmental harm attributed to the technology and its owners.(9) Society has become far less forgiving of mishaps related to innovation. It is no longer necessary to prove negligence to heavily penalize technology developers. Newspapers are full of examples. Alliances spread the risks as well as the returns from developing, validating, and introducing new technology.

Strategic alliances may also help break through barriers in foreign markets. Public policies set by local and national governmental agencies may deny local market access to non-domestic companies. The use of local technical capability can be one avenue through legal restrictions. In addition to technology, the local partner may provide better information and intelligence about customers and competitors.

Market relationships govern the continued success of a firm. Sometimes a company must help preserve the corporate health and technical proficiency of critical customers or suppliers. Imposing more stringent specifications and higher quality on a supplier's contribution may be fundamental to a company's ability to deliver customer-demanded product quality. Commitment to and dependence upon valued and necessary strategic partners may be strengthened through collaborations that lead to higher total quality.

The list of reasons and benefits for alliances is long. Other factors to be considered include management training, improved technology transfer, standardization of products and processes, and the use or sale of byproducts. The theme underscoring all of the suggestions is that management has to "think" alliance to understand the multitude of opportunities and resulting benefits. At the same time, though, care must be taken when entering into the complex and often confusing experience of technical cooperative projects.

Risks

The fear of "loss" causes many managers to shy away from forming partnerships outside the firm. That is, managers may fear that the firm will lose its competitive distinction which could be fatal. This must be guarded against in any strategic alliance. Other losses, however, tend to be self-inflicted through poor management. These include the loss of control and flexibility that result from the project's non-traditional structure and interdependencies. An expanded bureaucracy can cause a loss of efficiency, as response time to requests and problems slows. Incompatibilities of culture and strategic objectives not only create barriers to progress but can cause a loss of identity. Alliances that contain more competitiveness than cooperation are doomed to failure. Joint project management and participants must be wary of "hidden agendas." A partner may enter the alliance under the pretext of creating a commercial technology, when in reality the alliance was formed to fund self-serving research interests. If the research is irrelevant to the strategic objectives of one of the partners, or there is a feeling of uncertainty about the output value, the alliance will probably fail.

In some cases, the research foundation for the technology is a success but the development of a commercial product falters. Managers must prevent coalitions from focusing too much on the research phase without adequate consideration of the development phase that leads to commercialization. If the sole intention of the alliance is to advance the technology's fundamentals and not to apply innovations to marketable products, this must clearly be stated as a priority. If not, misunderstanding and conflicts will occur.

Potential Alliance Partners

When analyzing the external environment during the strategic planning cycle, R&D managers should consider four primary candidates for technical collaboration: customers, suppliers, competitors, and complementary firms. The firm's relationship to potential partners determines the type of technology alliance to form. The growing interdependence among these vital business connections provides added incentive for using alliances to strengthen these links.

Customer Alliances

Recognising the importance of the customer is the foundation of successful business strategies. Looking to possible customer needs for technological advancements may provide great opportunities for R&D. In some cases, the customer's survival may depend on these innovations. Certainly, it is in the firm's best interest that its customers prosper. An R&D alliance with customers may provide an advantage over other suppliers and help garner more of the customer's business. Customer alliances permit longer time horizons for the participants to plan and grow together. By serving the customer's technology needs, a firm benefits by extending its own technical frontiers.

An example of this type of alliance is found in the health care field, where cost containment is a paramount concern. PCS, Inc., a unit of McKesson Corporation, has signed a contract with four health insurance companies to create an electronic information network to assist in their cost containment efforts. PCS is expected to design a system that is similar to a bank credit card system. After receiving a service, patients will present their health-care card to their provider. The clerk will process the transaction like a credit card purchase, except the charge will go to the insurance company. This will eliminate the need for either patient or provider to complete insurance forms.

It is natural to consider customers as external to the firm. For a corporate R&D program, however, the potentially more important customers may be internal. Formal alliances with operations and marketing can provide significant direction and guidance to profitable new ventures. Marketing can furnish a deep understanding of customer needs and market demands. Operations have the knowledge and experience to cost effectively produce and deliver the product to market. Research and development supplies the creativity and technical skills to satisfy the market ahead of the firm's competitors. Failure to capitalize on the synergism in interdepartmental cooperation can lead to over-designed, over-priced, obsolete, or radically-advanced products with little or no consumer value. For many years, the U.S. auto industry epitomized this insular mentality.

Supplier Alliances

Just as a customer provides revenues to the firm, suppliers may represent the bulk of the costs. Furthermore, the suppliers' R&D investment can be as substantial as the firm's. Because the company's products and processes depend on healthy suppliers, R&D management must look backward through its supply channels when planning research strategies. For example, IBM's successful (though slow) PC product development depended on operating system software developed by Microsoft. Microsoft and IBM have gone on to develop other interdependent products. For example, Windows was developed for IBM PCs as Apple's Macintosh machine made inroads in IBM's office market. Microsoft does not play favourites; it has developed software for Apple computers since 1984.

Establishing suppliers as partners is generally a win-win situation. Alliances geared toward reducing supplier costs or improving the quality supplied can greatly affect the attractiveness of the firm's own products and services to its customers. Moreover, concern for the welfare of valued suppliers, who are committed to the firm's objectives and who have demonstrated the ability to perform, further strengthens the business relationship.

Competitor Alliances

The decision to form an alliance with a competitor is seemingly at odds with conventional wisdom, but competitors must be considered as potential strategic resources. In order to identify prospective competitor alliances, the manager must form a dispassionate view of the firm and its competition. The competitive scene should be described "in terms of fine structure that underlies the broad picture."

The key to deriving mutual benefit from a competitor alliance is to fully understand the major difference between the two firms. Planning and operating must be done in the context of this differentiation; otherwise the firm will lose its competitive distinction. With this understanding, a cautious and calculated position will be taken in specifying and limiting the nature of cooperative agreements. Care must also be exercised in setting limits to avoid breaking anti-trust laws.

The attitude and viewpoint of the competitor partners must also be honest and forthright. If the purpose of the alliance is just to avoid investment or development effort, or to gain access to competitor technology or market share, tremendous problems will ensue. However, if the alliance is a strategic move for jointly stronger competitive positions, for shorter response time to market changes, or due to over-demand on or lack of internal resources, then mutual benefits are possible. This will still require sustained effort and investment from both sides. The Advanced Micro Devices and Fujitsu "flash memory" chip alliance is a good example of this strategy. The firms have agreed to manufacture these chips jointly and in addition will buy stock from each other. Another example is the IBM/Toshiba/Siemens alliance.

Complementary Alliances

Complementary alliances exist when two firms possess similar technology but different product lines. Many basic technologies span several industries (e.g., superconductivity in consumer electronics, scientific instruments, electric components and motors). A single technology may be implemented differently by firms with different products in various markets. The scope of their R&D in the field may also be quite varied. A coalition of their energies and resources may yield much greater advancement of the overall technology than the sum of their individual efforts. This kind of technology coalition may be classified as a "vertical" alliance. Because the value added to each firm is difficult to quantify at the onset of the joint effort, specific objectives associated with each firm's competitive strategies are difficult to define. The closer the alliance members are to exploiting any resultant innovation, the more definitive the objectives must be.

Complementary alliances also exist between large and small entrepreneurial firms. Cooperative relationships between large and small firms in high technology can bring to market new innovations that neither firm alone could have accomplished. This synergy results from the small, entrepreneurial company offering innovative technology while the larger firm supplies the necessary marketing resources. In either case, combining complementary strengths enhances each firm's competitive position and financial performance above what individual paths could have achieved. The relationship is strengthened by the degree to which their competitive advantages differ.

Chevron Chemical, a unit of Chevron Corp., and Ecogen Inc. have formed a complementary alliance that fits both types. Ecogen is a small agricultural biotechnology firm with $6.9 million in revenues. Chevron Chemical operates in the agricultural chemical field and has about 40% of the approximately $100 million retail biopesticide market. Their products are, in general, complements rather than substitutes. The alliance will permit Ecogen to market its biotech products to retail customers through the distribution channels developed by Chevron Chemical for its Ortho brand products. Both will be marketed under Orthoganic's environmentally compatible line of pesticides and fertilizers. Presently, this alliance serves both firms well: Ecogen gains access to a new market and Chevron gains a broadened product line. This type of alliance can produce unexpected results. What began as a cooperative project can quickly move toward competition among alliance members. Managers must exercise extreme care to prevent unplanned transfers of competitive technologies as the unexpected competitive environment develops. A change in attitude and formal arrangements regarding the relationship of the partners is required. A competitor alliance may be more appropriate. An alternative, depending upon how the linkage consolidates, might be to form a joint venture company or discuss a merger.

Facilitating Alliances

It is not unusual for agents outside the firm's product or market environment to foster alliances. Actions by government and academic institutions can encourage cooperative efforts. Two examples stand out. The Federal Trade Commission is sponsoring a competition for the best high definition television system. In 1987, when the competition started, the U.S. was not expected to be a player in HDTV. However, with governmental encouragement, two American groups are presently leading in the race: Zenith/American Telephone and Telegraph and General Instrument/Massachusetts Institute of Technology. The FCC expects to choose the standard by the end of 1993, and U.S. television viewers are expected to have an American-developed HDTV system by 1996. This is an extremely short time for such a complicated R&D project to go from the drawing board to the home. The relationships have been greatly strengthened by the willingness of each partner to share technology. A second example is a response to California's requirement that auto makers sell 40,000 emission-free cars a year by 1998 and 200,000 by 2003. Calstart, a consortium of 43 organisations ranging from electric utilities to aerospace companies to academic institutions, was formed to meet the challenge. The consortium's success will depend on a willingness to share technology and to commit the resources. Apparently, there is progress. Calstart announced that systems and components for a prototype electric car will be displayed in auto shows around the world in 1993.

Research coalitions with universities might be considered another form of supplier alliance. Increasingly, the world's academic institutions are major suppliers of advanced technologies. About 60% of all U.S. basic research and a smaller, but still substantial, percentage of applied research is performed in U.S. universities. Because the transfer time from theory to application is rapidly shrinking, academic contributions are directly shaping the form of new product development.

Researchers in the academic communities can be in scarce supply. In order to affect the type and direction of university research, industry must participate with both money and personnel in collaborative efforts. In addition to influencing the course of research, joint participation has recruiting benefits. Firms can more readily evaluate prospective employees from the graduate ranks, and the students obtain an inside view of future employers.

The latter is extremely critical for recruitment in scientific and engineering disciplines where the number of annual graduates is limited and the competition to hire them is fierce.

Developing Successful Alliances

Studies of technology alliances have identified many issues which must be addressed to ensure optimal results. These issues have several key areas to consider: selecting the right partners; the quality of the relationships of those involved; the development of the project on which the alliance was founded; the detailed agreement specifically outlining individual and joint partner rights and responsibilities; and the management system employed to accomplish the project's goals.

Partner Selection

The first and foremost task in forging a successful technology alliance is to select the right partner or partners. Collaborating with the right partner satisfies several key requirements:

- ensuring benefits received by all participants outweigh the drawbacks;
- getting powerful and assertive top management dedicated to the joint project; and
- gaining project personnel with the competence and commitment to succeed.

To assure a harmonious relationship, the partners must be culturally compatible. A working relationship will not prosper without mutual respect for cultural differences or if either partner tries to dominate. A smaller partner typically wants to preserve substantial autonomy, primarily to retain innovative efficiency and flexibility. The demand for freedom from control may be the stopper for many large corporations. Experts are split as to optimal amounts of influence and intervention from the larger firm. Some advocate that the less the degree of control from the larger firm, the more probable the success of the alliance. Others insist that a hands-off policy by the large firm does not guarantee success and that the larger firm must be an active participant. Extensive prior contact is a critical aspect for adequate assessment of compatibility.

The firms must also offer comparable levels of scientific or engineering competence, otherwise the lack of professional respect and equity will undermine effective cooperation. Parity of technical competence is necessary, but not sufficient. Common or compatible design philosophies (methods of approaching problems) must also be sought or the project could be jeopardized with perpetual disagreement. Difficulties may arise if either partner feels poorly served by the alliance.

Synergistic benefits are also necessary for a successful alliance. All sides must gain. When synergy does not follow the alliance, dissolution is the likely result. Finally, there can be no hidden agendas. Strategic objectives must be compatible and in concert. Again, familiarity and effective communication are extremely helpful in avoiding misunderstandings and irreconcilable differences. When two or more of these elements are missing, difficulties are inevitable. For example, in 1982, SmithKlein, a proprietary and ethical drug firm, purchased Beckman Instruments, the leading producer of medical instrumentation.

Presumably, product and market synergy existed as well as a common commitment to R&D. In this alliance, both firms are more equal in size and have similar product lines. Furthermore, their product introduction schedule meshed. SmithKlein had no drugs in its pipeline due for release until 1993-95 while Beechman had many ready for introduction between 1989 and 1993 with none thereafter.

Quality of Relationships

Regardless of the type of alliance, cooperation cannot be mandated from above. The ultimate success or failure to achieve the intended targets resides with the people directly involved. Because these people are uniting from different organisations, the quality of their working relationships takes on greater importance than if they all belonged to the same work unit.

Thus, new working relationships must be formed in any joint endeavour. These relationships must allow project members to constructively confront, challenge, and compromise in a give-and-take manner. This builds mutual trust and promotes cooperative problem solving. To achieve mutual respect, all parties involved must be highly committed and competent. Experimental learning is desirable or even required and is particularly advantageous if the participants have prior knowledge of each other's capabilities and expectations.

Project Selection and Specification

Business agreements require formal contracts, and a technology alliance is no different. However, before contractual formalities progress very far, it is essential that the project be clearly defined and agreed upon by the project researchers. Definition must include the purpose, scope, objectives, and task descriptions. Establishing specific objectives will help those involved to form harmonious working relationships.

The project should not only be compatible with the desired results, but also narrowly focused on those results. As early as possible, precise and if possible quantified specifications should be issued showing the contributions and responsibilities of each firm and individual participants. The care exercised in project selection and specification will greatly reduce future complications from missed or unclear assignments.

Formal Agreement

The value of foresight when preparing formal agreements cannot be overstated. Many failed cooperative research efforts can be attributed to poorly drawn contractual procedures. For example, unexpected or unmanageable problems or disagreements that could delay or degrade the returns from the new technology may arise after the project begins. Mechanisms to deal with them constructively should be included in the agreement. Even after an R&D phase is successfully completed, there may be disagreements over usage rights and royalties. Aspects of property rights should be stated:

1) what innovations belong to whom;
2) how unanticipated innovations will be jointly shared; and, 3) how licensing rights will be handled.

In addition, revenue distribution should be decided before commercialization.

To facilitate constructive negotiations, several factors can be helpful. First of all, representation of all interested and involved organisations early in the contract negotiations can prevent otherwise predictable difficulties. Additionally, it is helpful if the negotiators are conversant in both the technology in question and in contract law.

A delicate balance must be struck in clearly defining project objectives, member responsibilities and commitments, and resources each partner is to contribute. On the one hand, the agreement should provide researchers flexibility to adjust and adapt to unforeseen hurdles or program changes. On the other, the agreement should not be so general that it causes confusion in direction or purpose or creates ambiguities over rights and privileges.

Ideally, provisions should be included to effectively resolve potential obstacles that cannot be predicted accurately. In the final analysis, all participants should feel they are deriving benefit commensurate with their contribution.

Other Success Ingredients

The decision making process should be decentralized to assure: a sense of parity of responsibilities and authority; a knowledge of the problems and people involved; better understanding of alternative solutions; and closer communication links. As a corollary, management should have dispute resolution mechanisms in place. Further, a conscious effort will probably be necessary to instigate dispute procedures, because the traditional management structure is inappropriate in a multiorganisational group. These mechanisms cannot be so stringent for the sake of expediency that they force improper resolution of controversial problems. Decisions on these issues may take longer, but the extra time may reduce mistakes. Management from all partners must be integrated into the problem solving process.

Second guessing of colleague intentions, actions, or decisions should not be tolerated. Professional courtesy and respect is lost when others take it upon themselves to make judgmental calls without consulting the entire research team. To do so breaks down the working relationships within the group and violates the reasons for initiating the cooperative project in the first place. Joint management should anticipate and avert increased specialisation among partners. Specialisation weakens the ability of the alliance to complete the entire project. It also may create undesirable levels of independence among the partners, because each "owns" a particular piece of the process. The alliance must preserve joint ownership and control of all aspects leading to the product. Geographic proximity is extremely important to any type of cooperative project. Continual contact promotes communication and creates understanding and trust through personal relationships. Close proximity can reduce product development time. In general, it promotes a real sensitivity to direct interaction.

Managing the Alliance

Because the alliance is collegial rather than hierarchial, most traditional managers must rethink their approach to alliances. The alliance manager must accept that, because of the interdependence among the participants, constraints will be placed on his or her courses of action. One constraint is the need for the coexistence of different cultures without one becoming dominant. The managers must nurture the network of contacts that form the foundation of mutual trust and cooperation.

Managers representing partnering firms must fully understand the critical technologies to be developed and shared and the core technologies of his or her own firm that will remain separate. It is essential that internal proprietary technologies be protected in order to preserve a firm's competitive edge. Yet, the alliance must gain from pre-existing technology know-how. Managers must share appropriately if the alliance is to enhance competitive advantage, not reduce it.

Milestones and checkpoints should be set, and management should closely scrutinize them. Control over the project's development and decisions can be maintained if there is total cooperation among project members and management. Timely information exchange at critical points is a strong catalyst for cooperation and openness.

Conflict is a fact of life and can be a powerful tool for cooperative effort if managed constructively. The negative aspects of mismanaging conflict can be detrimental to teamwork and may lead to the demise of the alliance. Managing conflict correctly can generate better ideas and new approaches through the examination of alternative viewpoints. Long-standing problems may surface and be tackled. Project members forced to clarify their positions will create a tension which can stimulate interest and creativity. The greatest benefit is the increased cohesiveness within the alliance after the problem is resolved.

A technology alliance can be a powerful weapon in the strategic manager's arsenal of options. For firms lacking sufficient funds or internal expertise to achieve all their goals, a strategic technology alliance may be the only option available. The strategic technology alliance can also provide a solution to a problem by meeting objectives already set. Additionally, alliances should be considered as part of the planning cycle to identify opportunities in the firm's business environment. Alliances with users, suppliers, and other organisations (including competitors) may strengthen current market positions or identify new ventures. However, management must carefully assess the risks and benefits. Then, once it is clear an alliance is appropriate, all partners must work together to make it succeed. Alliances are strategic competitive tools if the firm wishes to continue to compete and lead, management must begin to "think alliance" and learn to reach out effectively. The global economy will not permit many firms to remain insulated and survive.

Information Systems for Managerial Decision Support Systems (DSS)

Management must accept ultimate responsibility for the success or failure of a DRP; therefore, communicating thc importance of a DRP to management is the first step in establishing such a plan. The audit committee/board of directors and the comptroller are the most critical levels of management that need to support a DRP. However, the entire company needs to be involved with the DRP process. Upper management needs to provide the strategic direction and financial resources for the DRP. Middle management is responsible for facilitating the coordination and effective execution of the plan. Lower management should understand the need for a DRP and provide insights into plan improvements and efficient execution.

There are two sources for initiating communication to management about the importance of a DRP. One source is external, the external auditor and/or computer consulting companies that provide services to the company. The external auditor is expected to provide advice concerning internal control structure weaknesses. An inability to recapture information, whether from a disaster or general system vulnerability, is considered an internal control weakness. The other source for initiating communication is internal, an employee of the company. This employee could be from one of several departments, e.g., internal audit, information systems, comptroller or a specific individual concerned about information security.

When communicating the importance of a DRP several issues should be emphasized to management. The effect of even a short-term interruption in business on profits and cash flows should be stressed. Specific examples of disasters are the fire that destroyed the corporate headquarters of Steinberg, Inc., a Canadian company with $4.5 billion in retail sales, the bombing of the Word Trade Centre, the Con Edison fire and the Los Angeles and San Francisco earthquakes. These examples illustrate how quickly a company can be crippled by an uncontrollable event.

A competitive disadvantage for the company during even the shortest business interruption is that competitors will have the opportunity to sell to the company's customers. Management should be reminded of how detrimental this disadvantage will be to the company in both the short and the long term. Also, the potential exists for stockholder lawsuits for gross negligence related to safeguarding the company's assets. The underlying theme of these issues is that there are substantial costs associated with the risk of not implementing a DRP. The cost of not having a plan in place can be severe enough to cause bankruptcy. One question should be asked of upper management: "Are you willing to assume such risks when a relatively small investment of time and money can provide a solution?" The process of establishing a DRP is not costly and is successfully being performed by numerous companies, such as Steinberg, Inc.

When establishing a DRP, management should pay special attention to assigning responsibility to specific individuals. One approach is to hire a full time employee who is designated the disaster recovery person. Whether a full time employee is or is not hired, the responsibilities for a company's DRP should be distributed among several employees. The best approach would be to designate a disaster recovery team, with one member specified as team leader and another as assistant team leader. The disaster recovery team should be responsible for establishing and documenting a DRP with the specific tasks associated with the plan divided among employees. Once a disaster occurs, one individual should be responsible for beginning the recovery process. This individual is responsible for communicating with employees, customers and suppliers. The communication process should include issues like how to contact the company (new phone numbers), where the temporary office is located and how payroll obligations will be satisfied.

Another employee should be assigned responsibility for maintaining the currency of the DRP. Tasks associated with maintaining currency of the plan include identification of the following items: critical data files, key employees, location of back-up sites and how to replace hardware and software. As part of the process of identifying key employees, it is crucial to identify and document the specific duties of these employees along with developing a hierarchical list of employees to take over these specific duties.

An employee should be specifically assigned responsibility for backing up critical data files at specified intervals and securing these data files at an off-site location. Data represent the core function of an information system. Hence, the importance of this task cannot be overstated. Additionally, it is crucial that written documentation of all components of the DRP be maintained at locations outside the company. Homes of the DRP team members offer possible options for storing copies of the DRP.

Operational Information Management

Initially, the database was created on CSIRONET by dispatch of coding forms to CSIRO from AMF for paper tape data entry. Later data entry took place directly from AMF, and from 1982 this was managed through a host DEC PDP11/44 minicomputer for validation, then storage on a Cyber76 on CSIRONET in Canberra. The thesaurus was transferred to the PDP host. Software support was provided by the CILES System Development Group. The live database was updated monthly on CSIRONET. From 1980, quarterly updates were also produced for Ausinet where they were mounted after conversion to STAIRS with software developed by ACI Computer Services. Full document backup (or referral for unpublished documents) was provided by AMF. In 1987 AESIS was relocated from CSIRONET to CLIRS as part of its Australian Resources Industry Database concept.

Other operational aspects concerned the continuing maintenance of the thesaurus terminology and production of new editions, as well as the production of other titles that were structured along the same lines as AESIS. For example, Earth Science and Related Information Selected Annotated Titles (ESRISAT) selectively covered earth sciences serial publications received by the AMF and South Australian Department of Mines and Energy libraries and State Library of South Australia. Seven indexes: subject, locality, author, map sheet, mine/deposit/well/name, stratigraphic and serial title, were created for the monthly service which also had semi-annual cumulations. These were the same indexes as for AESIS, and the material included incorporated AESIS updates along with library acquisitions.

Document delivery costs estimated at $5 per request excluding requester's cost for normal (comparing with quoted national figures of $5.56 and lending of $3.72) although 90% are about $3.60 are close to the national figure and 10% are about 4-5 times that.

There were concerted efforts to develop STI services in Australia during the 1960s and 70s within a public information policy framework. However, although these efforts led to greater awareness of the issues, national development lacked a strategy that stakeholders could follow to avoid gaps in service and duplication. This situation was exacerbated by funding constraints. However, a rapidly developing computing and communications environment coupled with the efforts of some visionaries working independently in different agencies, saw to it that the country was comparatively well-serviced using a combination of international and local services.

One of the agencies in the vanguard was the AMF, whose AESIS service provides the focus for the case study. The initial success of AESIS can, in no small part, be attributed to the acuity of its management, and it provided a prototype for similar Australian services. Despite the demise of its harboring organisation, the quality of the database has seen it revived in a different context for the petroleum and exploration industry. However, its continuation will happen effectively only by application of the collaborative principles that contributed to is original success.

This case study approach aspires to test whether a particular service is carried out according to the tenets of a domain-based understanding of information management. This requires attention to be paid to planning and strategy through administrative, analytical and operational aspects. Administrative aspects may have to take into account public policy, and analytical aspects include user needs and systems analysis. The AMF was found to be conscious of the need for consideration of each of these domains, though the elements of the domains were not articulated as such by the enterprise itself at the time of development.

The three domains have proved in be useful in this case for conceptualizing an application of information management. They represent an approach by Middleton (2002) that endeavours to illustrate how information management reconciles information science principles. Therefore if such understanding can be applied in similar cases, it should prove useful for the planning and development of services. This study is limited by focusing on a single case, by limited recourse to historical records, and by examining the service at a time when it was no longer operational in the same way as during the period under consideration. However, subsequent case analysis of similar STI services is showing promise in confirming the analytical framework. Whether the analysis can be extended to information services in general is problematical. However it seems to provide a useful understanding at least in this constrained area, of those factors that need to be addressed to make such a service work well according to tenets of the field.

Development, Management and Control of Information Systems

Advances in information technology and perceived dissatisfaction with MIS performance is leading users to take over their own systems development work. This does not mean an end to the MIS department, but a staff rather than line responsibility will be required as users become the dominant developer of information systems. For a successful transition, HRD will be expected to operate as a change agent helping both groups adjusts to their new roles. The introduction of microcomputers into the workplace during the 1980's ushered in a new era which is having a profound effect on organisations.

More specifically, users are taking greater control for systems development in their organisation. This change requires 1) user departments to prepare for new responsibilities and 2) the Management Information Systems (MIS) department to adapt to a new role and purpose within the organisation. Furthermore, the Human Resource Department (HRD) needs to help manage the conversion from an MIS dominated to a user controlled environment. This chapter identifies the causes that are precipitating a change in information systems responsibilities and describes necessary modifications in roles, for user departments and the centralized MIS department, as a consequence of the change. Recommendations on how to prepare for the transformation that is coming are also provided.

Technological Factors

The early dominance of the centralized MIS department (CMIS) was a result of the high cost of the technology. Expensive computers, and the need to have a CMIS employee program the computer, centralized computing in one department where the mainframe was the centrepiece of the operation. In the 1970s the development of smaller computer syst mputing work, predictions were made that it would not be long before users would be doing most of their own systems development work. As the decade closed, reports began to appear to support that forecast.

One of the obvious changes that end user independence precipitates is a decentralization of the information system function. The role of CMIS within the organisation begins to change as decentralization expands, and faced with a situation where they are less likely to control the information systems environment, a coordination role for CMIS becomes a viable alternative. In addition to the affordable price of hardware, which places computers within the reach of many user departments, the efficiency of PC's relative to mainframes is an important consideration for a cost conscience government. A popular measure of computer performance is MIPS (million on instructions per second) and the cost per MIPS is measurably less with PC's compared to mainframes. This economic consideration also favours an acceleration in user departments justifying their own hardware.

However, it is not just new hardware (e.g. PC's and local area networks) that encourage users to take greater control over their information processing. More user friendly and powerful software also provides members of the functional departments with the potential to produce impressive results with their PC's.

The trends described above are clearly reflected in the evolution of computing in state government where the proportion of small computer systems rose consistently throughout the 1980's and there was a rapid growth in PC's in latter part of the decade. In addition, departmental computing continues to expand and state agencies are moving applications off the central mainframe to minicomputers. Nonetheless, mainframe operations are not diminishing since new mainframe demand is being generated by departmental networks. States are also anticipating further technological and managerial changes which are indicative of a strong user's orientation - greater use of powerful small computer systems, growth in user computing and an increase in computer networking and data sharing.

Organisational Factors

In addition to the technological advances described in the previous section, organisational factors have weakened the control of CMIS over information systems. The primary issue is the performance of CMIS as perceived by users. A survey by The Partnership for Research in Information Systems Management or PRISM illustrates the type of disenchantment users experience with CMIS.

The survey found that 75% of users, who had acquired their own systems, cite unsatisfactory performance by CMIS as the most important factor in wanting their own system. These users also justify their independent systems on the grounds that they can produce systems more quickly and better tailored to their needs than CMIS. The importance of flexibility and access are additional arguments used by users to obtain their own systems.

Furthermore, functional managers perceive CMIS as unresponsive to their needs. A survey of user managers found that 40 percent of CMIS installed systems, which are used for important departmental activities, were judged to be inconvenient, inflexible or incomplete by the managers. What seems clear is that not only are the tools for user independence available, but the dissatisfaction level among users makes the situation more susceptible to change. There are signs that performance problems are recognised by state governments. High costs, missed programming schedules, costs overruns, inflexibility of programs and programming backlogs are listed as the 5 top problems in state CMIS organisations. This has been a consistent concern during the 1980s.

On a pragmatic level, a shift in the role CMIS departments play in their respective organisations show that the environment in state governments is changing. There is a steady movement away from a controlling function; CMIS is now more likely to operate as a coordinator of information systems.

Adding to the Duties of User Departments

The most pronounced change for user departments is the opportunity to develop their own information systems. As Allen observes, users are in a better position to evaluate and acquire their own information systems than CMIS. Users know their departmental operations far better than CMIS, are in the best situation to judge the importance and priority of needed systems and are in a better position to react quickly to outside pressures. If users can obtain hardware and software independent of CMIS, Dearden concludes that there is no reason why they should not exercise complete control over their information systems.

From an organisational standpoint, departmental independence is also consistent with a stronger authority - responsibility relationship. Giving user department's greater freedom to choose and operate their own information systems means they also must accept the responsibility for these new functions. In order to fulfill their new obligations several basic requirements need to be satisfied. A minimal requirement is for user departments to be headed by managers who are computer literate. Being computer literate does not require that the manager possess specific skills such as programming or the ability to use PC application software such as word processing or a spreadsheet. However it does mean the manager has an appreciation for how computers can be used to accomplish departmental tasks, how they affect individual operations and what their impact is on individuals.

Specific information systems functions will fall to departmental staff with more specialised training in the discipline - tasks familiar to CMIS such as planning strategic departmental information systems, determining information needs, obtaining hardware and software within guidelines established by CMIS, assessing whether systems are in conformity with regulations and policies, and evaluating the exposure of risk from information systems failure. Fortunately, the execution of these responsibilities does not have to be done in a vacuum. With the help of HRD, user departments can utilise the experience and competence of CMIS in the exercise of these assignments.

Restructuring the Role of CMIS

A change in CMIS duties appears inevitable. Although the change will be gradual, it is felt that CMIS departments, as they are currently constituted, will be substantially modified. This does not mean the demise of CMIS, but it does mean a restructuring of the role CMIS plays in an organisation.

Planning and Standards Development

The greater the dispersed environment becomes, the greater the need for centralized planning and control. CMIS is the most reasonable place to locate the information planning and control responsibility for the organisation as a whole. In particular, hardware and software standards need to be formulated. Standards for data management practices, privacy issues and security concerns also need to be established.

Technical Leadership

The flow of new technology provides an opportunity for organisations to be more efficient and effective. Although government in general is not on the leading edge of computer technology, their size and function provides large dividends when adopting cost efficient measures.

Technology surveillance is considered an important role for CMIS in its new charter. The background and experience of the professional CMIS staff is particularly well suited to perform the technology tracking function within organisations. In addition, a major challenge for CMIS will be to facilitate the transfer of user developed systems as an organisational resource.

User Support

The change in user department/CMIS roles places a greater responsibility on CMIS in the area of user support. One of the best vehicles available to CMIS to support users is through an information centre, i.e. a facility with hardware, software, consulting and educational capabilities to teach and support the user. An additional responsibility for a central unit is to form user groups and serve such groups in a coordinating fashion.

The establishment and maintenance of local area networks represents another area for user support. As PCs become more numerous and powerful, computer networks become the preferred means of departmental communications.

Nevertheless, common databases and connections between departmental networks must be planned, created and maintained. Its unlikely that any department other than CMIS would choose to take on that responsibility.

Computer Operations

A viable data centre operating the mainframe and ancillary equipment will remain an integral part of CMIS. However, Zachman notes that there is a controversy over the future of the mainframe and visualizes its eventual replacement by PCs. Martin points out, that in the future, PCs will be capable of running the applications that are now run on the mainframe. However, for the immediate future, mainframes will be part of the information architecture of large organisations.

Systems Development and Maintenance

The point has already been made that user departments will be able to satisfy most of their own systems development needs. Nevertheless, systems that exceed the competencies of a user department, or effect the entire organisation, still have to be produced. When these situations occur, CMIS may be called upon to develop the system. However, another viable alternative is to have the system developed by outside software specialists. This alternative is seen as an option which may be particularly interesting to user departments.

Although there will be a decrease in systems development work, the manpower required for systems maintenance is expected to increase. These authors also believe that CMIS may assume responsibility for the maintenance of user-developed systems. On the one hand this translates into additional work for CMIS. On the other hand it creates a personnel problem within CMIS since systems maintenance is perceived by CMIS professionals as de-motivating, low level work.

The Situation in State Governments

The growing investment in information technology has led state governments to place greater importance on information resource management. Data compiled by NASIS support this appraisal. Most states have drawn up an overan statewide plan for information systems. In addition, 92 percent have developed formal policies governing the procurement of electronic data processing equipment. On the other hand, plans for distributed data processing, data sharing, and office automation are only available in about 50 percent of the states. Furthermore, progress in preparing privacy and security plans within state governments are minimal.

The growing importance of user training in state governments is reflected in the percentage of CMIS resources devoted to this function. The percentage of users trained grew from 26% in 1980 to 50% in 1988. There is a suggestion that the systems development emphasis is already easing in state governments. In 1988, the latest year for which data are available, programming backlog was rated as the fifth most serious problem facing CMIS management. In the five previous years programming backlog was rated as one of the top two problems for CMIS management.

The future role for CMIS is one in which more attention is focused on planning, consulting and monitoring the technology. The current major effort devoted to systems development is deemphasized and computer operations takes on a different look with a larger proportion of resources devoted to setting up and running user networks.

Managing the Transition

The transition to a user dominated environment will not be easy and many of the problems that need to be faced involve personnel issues. For example, the PRISM survey revealed that in some CMIS organisations there is a "sinking ship" attitude. As users take control of their own systems and MIS decentralization proceeds, employees in CMIS become less sure of their futures. They seem to lack understanding about their long term career options and thus, it is not surprising to find that CMIS professionals are being urged to take more responsibility for their career development.

From the user's perspective, it is important to consider the type of MIS skills needed in a departmental setting. CMIS systems developers rely on technical computer science skills - they are specialists in a single discipline. In a user department, these people need to rely on other attributes, and in particular, a familiarity with the professional orientation of the department. They need to acquire a working vocabulary for the new area, understand the operational procedures, and become adept at the problem solving techniques used by the department. In addition these individuals must be seen as allies, who are adept as consultants. The control and enforcement orientation of the past CMIS ethic needs to be replaced by an ability to elicit creativity and a talent in showing how information technology is used to benefit the department. The value of the departmental MIS expert will depend on skill in communicating and presenting ideas.

A revision of department job descriptions will be necessary since current descriptions reflect the way computing was done in a CMIS controlled environment. In addition to updating the outdated descriptions, job ratings require attention as well. User departments will be staffed with people, possessing not only technical skills in their functional area of responsibility, but with a proficiency in a second discipline, MIS. The greater breath of these jobs will need to be evaluated and adjustments made. Caudle and Marchand report that there is a move by states to reclassify job titles to better represent the work of information technology and resources. On the other hand they also note that there remains a considerable lag in the salaries of state MIS professional compared to those employed in the private sector.

Managerial issues also need to be addressed including overcoming resistance on the part of CMIS to the decentralization of the MIS function and changing traditionally defined MIS jobs. As user departments assume responsibility for MIS professionals, it is appropriate to ask whether MIS personnel can be handled the same way non-MIS personnel are supervised or should departmental administrators be prepared to modify their management style? Couger and Zawacki report differences between MIS and non-MIS people. Compared to workers in other disciplines, these authors find MIS personnel are less satisfied with the supervision they receive, have lower social needs, and possess a stronger desire for personal growth and development.

However, in a study by Ferratt and Short, no differences were found between MIS and non-MIS employees in respect to motivational patterns. The motivational patterns are similar for MIS and non-MIS personnel within each of three occupational groups (Clerical, technical and managerial). The authors believe that their results need not be interpreted as a contradiction of the Couger and Zawacki findings. For example, they note that if MIS people have fewer social needs they simply may not seek out interactions. However, an environment that is friendly and supportive, serves as motivation for productive work behaviour among MIS people. Hence it is important for departmental managers to intervene in order to produce the desirable social atmosphere since MIS people will be less likely to seek out such situations on their own.

The HRD Role

The role of HRD, as an organisation makes the conversion from a CMIS to a user dominated environment, is essentially that of a change agent. Consequently, an ideal background for an HRD representative involved in the transition is organisational development.

To initiate the process, HRD needs to communicate with CMIS and user departments to ascertain the extent to which the MIS role is changing. The main players in developing and implementing an action plan must be management representatives from these departments, and it needs to be understood that HRD's role is to facilitate the process and provide the necessary support functions. In addition, HRD must be sure senior management understands the situation and that all involved departments, as well as senior management, agree that it is in the best interest of the organisation to move to a new MIS method of operation.

A working group with representation from all involved departments and HRD create, problem solve and manage an action plan. It is also important for the departments to be in communication with their staff regarding the reason for and nature of the action plan. Although plans will vary depending on local circumstances, a number of essential elements include:

1) Determining the type and number of MIS positions required for each Department.
2) Creating job descriptions and a salary structure for new positions.
3) Matching current staffing versus needs and identifying areas of under and over staffing.
4) Developing training and development plans. Beyond the technical training, a plan to "socialize" individuals who transfer to a new department need to be considered. As the program evolves HRD should also be sensitive to the need to hold team building exercises within a department when appropriate.
5) Discussing with current staff their interest in a career change. In many cases individuals lack an understanding of the career choices within their present position and, even if they choose not to take on new responsibilities, they will profit from an understanding of their current options. Individuals opting for a career change need to be fully informed regarding the training needed and sources for the training (e.g., in-house or through outside organisations).
6) Instituting and monitoring individual plans. No matter how carefully thought out a plan may be, it does not guarantee success. HRD should retain contact with transferred individuals and depending on circumstances, HRD may wish to sponsor workshops which address problems or successful outcomes.

For years the computer caused job redesign and retraining of vast numbers of personnel outside of CMIS. As the information age matures it is now many of the professionals in CMIS who must learn new skills and alter their career paths and goals. It is clear that in state governments the transition from a CMIS controlled environment to a user dominated environment is occurring. Many states recognise the need for improved information systems planning and the facilitation of user training, but much still needs to be accomplished. Furthermore, change is never easy and is often painful. Yet the transition can be made with less hardship if the reality of change is accepted and if steps are taken to alleviate some of the most obvious problems. HRD needs to play an important role in the transition. As Bennis pointed out, HRD should be the catalyst for organisational change and adaptability. An early step in this process is for HRD to join forces with CMIS and the user departments. In a joint enterprise HRD can be instrumental in identifying career paths and opportunities for both CMIS professionals and user personnel. With a plan for the development and deployment of information systems personnel, the potential traumatic transition into the future can be a more controlled and rewarding process.

The Origins of Personnel Management/HRM

From the earliest times in Egypt and Babylon, training in craft skills was organised to maintain an adequate supply of craft workers. By the thirteenth century, craft training had become popular in Western Europe. Some authors suggest that the history of HRM can be traced to England, where masons, carpenters, leather workers, and other craftspeople organised themselves into guilds. Craft guilds supervised quality and methods of production and regulated conditions of employment for each occupation. The craft guilds were controlled by the master crafts worker, and the recruit entered after a period of training as an apprentice. The craft system was best suited to domestic industry, which the master operated on his own premises, with his assistants residing and working in the same house. The guilds used their unity to improve their work conditions.

The field of HRM further developed with the arrival of the Industrial Revolution in the latter part of the eighteenth century, which laid the basis for a new and complex industrial society. The Industrial Revolution began with the substitution of steam power and machinery for time-consuming hand labor. Working conditions, social patterns, and the division of labor were significantly altered. A new kind of employee—a boss, who wasn't necessarily the owner, as had usually been the case in the past—became a power broker in the new factory system. With these changes also came a widening gap between workers and owners.

As early as the 1890s, in some companies, a few specialised personnel activities were grouped into larger departments. The Civil Service Commission, established by the Pendleton Act of 1883, has had major influence on the development of personnel or human resources management in the United States. Drawing many of its ideas from the British civil service system, the Pendleton Act established the use of competitive examinations for admission into public service; provided job security for public employees, including those who refused to engage in politics; prohibited political activity by the civil service; and encouraged a nonpartisan approach to employee selection. A commissioner was appointed to administer the act.

The major effect of the Pendleton Act was to foster employees' appointment and career development in federal service on the basis of performance. Over the years this law has stimulated progressive personnel policies in private organisations as well. For example, around 1890 the Civil Service Commission was developing the forerunners of general intelligence tests and trade tests that became popular in private industry. In response to the growth of trade unionism at the turn of the century, a few organisations—created the position of "welfare secretary." Welfare secretaries were supposed to assist workers by suggesting improvements in working conditions, housing, medical care, educational facilities, and recreation. These people, who were the forerunners of today's personnel or human resources management directors, acted as a buffer between the organisation and its employees. The B.F. Goodrich Company developed the first employment department in 1900, but its responsibilities consisted only of hiring. In 1902 the National Cash Register Company established the first comprehensive labor department responsible for wage administration, grievances, employment and working conditions, health conditions, record keeping, and worker improvement.

Scientific management and welfare work represent two concurrent approaches that began in the nineteenth century and, along with industrial psychology, merged during the era of the world wars. Scientific management represented an effort to deal with inefficiencies in labor and management primarily through work methods, time and motion study, and specialisation. Industrial psychology represented the application of psychological principles toward increasing the ability of workers to perform efficiently and effectively.

Whereas scientific management focused on the job and efficiencies, industrial psychology focused on the worker and individual differences. The maximum well-being of the worker was the focus of industrial psychology. In 1902 Maryland became the first state to pass a workmen's compensation (now workers' compensation) law, requiring employers to pay workers for lost time and injuries resulting from occupational accidents. The law was subsequently declared unconstitutional. But in 1911 the U.S. Supreme Court upheld the workers' compensation laws of several other states, and from then on safety specialists became very common in industry. Industrial firms wanted to reduce claims against themselves, and they depended on the safety specialist to help ensure safer working conditions in their plants. As a parallel development, physicians were employed by some companies to ensure that employees would be assigned jobs suited to their physical qualifications.

The personnel department began to really emerge during the second decade of the twentieth century with drastic changes in technology, the growth of organisations, the rise of unions, and government concern and intervention concerning working people. In 1911 U.S. Steel created a Bureau of Safety, Sanitation, and Welfare. By 1918 International Harvester had established a Department of Industrial Relations, and Ford Motor Company had created a Sociological Department, which combined medical, welfare, safety, and legal aspects of employee relations.

In 1917 Standard Oil of New Jersey approved a plan that provided for regular conferences between labor and management. At the same time the company established a retirement income plan, substantial insurance benefits, a safety program, and a medical division. To coordinate many of the new programs, Standard Oil created a personnel and training department.

As organisations like Ford began to grow they started to hire their employees through newly created specialised units. Ford, for example, called this unit the "employment department." Although these units were initially created to hire employees, they soon began to help manage the existing workforce as well. This trend was influenced by a number of management books published between 1890 and 1912 in Great Britain and the United States. Moreover, the first comprehensive text in the field appeared in 1920—Tead and Metcalf's Personnel Administration.

Meanwhile, other developments, many taking place in other parts of the world, provided organisations with some of the tools they would need to better manage these employment processes. For example, in England the work of Charles Darwin popularized the ideas that individuals differed from each other in ways that were important. In France the work of Alfred Binet and Theophile Simon led to the development of the first intelligence tests, and during World War I several armies tried using these tests to better assign soldiers to jobs. These attempts at staffing continued in the private sector after the war, and by 1923 Personnel Management, a seminal book by Scott and Clothier, was already spelling out how to match a person's skills and aptitudes with the requirements of the job.

Another earlier contributor to HRM was called the human relations movement. Two Harvard researchers incorporated human factors into work. This movement began as a result of a series of studies conducted at the Hawthorne facility of Western Electric in Chicago between 1924 and 1933. The purpose of the studies was to determine the effects of illumination on workers and their output. The studies pointed out the importance of social interaction and the work group on output and satisfaction.

As businesses grew increasingly large, they began to create specialised units to cope with more than just their hiring needs. They also began to deal with government regulations and to provide a mechanism for dealing with behavioural issues. During the 1930s and 1940s these units gradually began to be called personnel departments (the word personnel was derived from an old French word that meant "persons"). They were usually set up as special, self-contained departments charged with the responsibility of hiring new workers and administering basic human resources activities like pay and benefits. The recognition that human resources needed to be managed and the creation of personnel departments also gave rise to a new type of management function—personnel management. The manager who ran the personnel department was called the personnel manager.

During this period, personnel management was concerned almost exclusively with hiring first-line employees—production workers, sales clerks, custodians, secretaries, blue-collar workers, unskilled labor, and other operating employees. Issues associated with hiring, developing, and promoting managers and executives did not surface until later.

Personnel management took another step forward in its evolution during World War II. Both the military and its major suppliers developed an interest in matching people with jobs. That is, organisations wanted to optimize the fit between the demands and requirements of the jobs that needed to be performed and the skills and interests of people available to perform them. Psychologists were consulted to help develop selection tests, for example, to assess individual skills, interests, and abilities. During the 1950s the wartime lessons were adapted for use in private industry. New, more sophisticated techniques were developed, especially in the area of testing, and organisations also began to experiment with more sophisticated reward and incentive systems. In addition, labor unions became more powerful and demanded a broader array of benefits for their members. At the same time, government legislation expanded and continued to add complexity to the job of the personnel manager.

Building on the findings of the Hawthorne studies, managers began to focus more and more attention on understanding human character of their employees. It was during this era, for example, that Abraham Maslow popularized his "hierarchy of human needs". Douglas McGregor's well-known Theory X and Theory Y framework also grew from the human relations movement. Briefly, Theory X rests on an essentially negative view of people. It assumes that they have little ambition, dislike work, want to avoid responsibility, and need to be closely directed to work effectively. Theory Y, on the other hand, rests on a positive view of people. It assumes they can exercise self-direction, accept responsibility, and consider work to be as natural as rest or play.

McGregor personally believed that Theory Y assumptions better captured the true nature of workers and should guide management practice. As a result, he argued that managers should free up their employees to unleash their full creative and productive potential. Building on the basic premise of the work of the human relations era, researchers such as Maslow and McGregor stressed that if managers could make their employees more satisfied and happier, they would work harder and be more productive. In more recent times, researchers and managers alike recognise that this viewpoint was overly simplistic and that both satisfaction and productivity are complex phenomenon that affect and are affected by many different things. Nonetheless, the increasing awareness of the importance of human behaviour during this period helped organisations to become more focused on managing their human resources. These organisations saw effective management of human resources as a means of potentially increasing productivity and, incidentally, as a way of slowing the growth of unionism, which was gaining increasing popularity.

The early history of personnel still obscures the importance of the HRM function to management. Until the 1960s, the personnel function was considered to be concerned with blue-collar or operating employees. It was viewed as a record-keeping unit that handed out 25-year tenure pins and coordinated the annual company picnic. Peter Drucker, a respected management scholar and consultant, made a statement about personnel management that reflected its blue-collar orientation. Drucker stated that the job of personnel was "partly a file clerk's job, partly a housekeeping job, partly a social worker's job, and partly firefighting, heading off union trouble.

As suggested in Drucker's comments, from its inception until the 1970s, personnel management was not seen as a particularly important or critical function in most business organisations. Although many other managers appreciated personnel management as a necessary vehicle for hiring bluecollar and operating employees, it was also seen primarily as a routine clerical and bookkeeping function—placing newspaper ads to recruit new employees, filling out paperwork on those employees after they were hired, and seeing that everyone got paid on time.

While other organisational units like finance, operations, and marketing grew in status and perceived importance, the personnel department of most organisations was generally relegated to the status of a "necessary evil" that had to be tolerated but which presumably contributed little to the success of the organisation. And personnel managers themselves were often stereotyped as individuals who could not succeed in other functional areas and who were assigned to personnel either because the organisation had nothing else they could do or as a signal that the individual was not a candidate for promotion to a higher-ranking position.

Contemporary Growth of HRM

Today, the HRM function is concerned with much more than simple filing, housekeeping, and record keeping. As time has passed, the role of HRM has changed dramatically and become much more important in most organisations. Perhaps the first major stimulus for this increase in importance was the passage in 1964 of the Civil Rights Act, which made it illegal for employers to consider such factors as gender, religion, race, skin colour, or national origin in making employment-related decisions. The 1964 Civil Rights Act, combined with several subsequent amendments, executive orders, and legal decisions, made the process of hiring and promoting employees within the organisation far more complex than ever before. Organisations quickly realised that those responsible for hiring and promoting employees needed to fully understand the legal context within which they functioned.

As HRM was becoming more important because of the increasingly complex legal environment, many managers were beginning to recognise that HRM had important strategic implications for the organisation as well. During the 1960s and 1970s, for example, international competition grew rapidly, and all organisations found it more important than ever to use their resources wisely and to capitalize on their full value. While managers were becoming increasingly concerned with ways to improve productivity and competitiveness, they also began to realise that workers needed to feel that their jobs were a source of personal satisfaction and growth. Successful organisations began to realise that they could maximize effectiveness and make work more meaningful and fulfilling for employees.

HRM departments in the 1960s and 1970s also had to become much more professional and more concerned about the legal ramifications of policies and practices. Also, organisations took a new look at employee involvement and quality of work as a result of concerns about the impact of automation and job design on worker productivity.

Given the shift in competitiveness top management in most organisations began to see that HRM practices and policies significantly affected their ability to formulate and implement strategy in any area and those other strategic decisions significantly affected the organisation's human resources as well. As a result, HRM began to be elevated to the same level of importance and status as other major functional areas of the organisation. The top HRM executive at most companies today has vice president or executive vice president status and is a fully contributing member of the organisation's executive committee—the executive body composed of key top managers that makes policy decisions and sets corporate strategy. The 1980s and 1990s brought some other changes to the HRM function as well. Many organisations found that they were not able to compete well in the new, global marketplace. Some of these organisations went out of business or were acquired by other, more successful organisations as a wave of mergers and acquisitions began in the United States.

After the merger of two organisations, there was often not the same need for as many employees, and many employees lost their jobs. Those organisations that could not compete and that were not acquired by some other company closed down, and yet more employees lost their jobs. Finally, those organisations that were struggling to be competitive often concluded that they could be more efficient with fewer employees and contributed to the era of downsizing, rightsizing, or reengineering. Regardless of what it was called, there were fewer and fewer jobs around.

During the 1980s, the strategic role of HRM became essential as organisations reduced staff, closed plants, or "restructured." The job of the HRM function was made more difficult but also had a more profound, direct effect on the HRM function with the increased employee displacement. As these organisations looked for new ways to be competitive and reduce costs, they looked for activities within the organisation that could be done more efficiently by outsiders. It also became somewhat common in some organisations to reduce the size of their HRM staffs and turn to outside help for specific projects. This outsourcing resulted in smaller HRM staffs and other organisational employees.

However, despite the fact that some HRM activities are being outsourced, the activities carried out by HRM are indeed growing in importance. Containing the costs of health care benefits also grew in importance. During the 1990s, organisational restructuring continued. The traditional HRM function began shifting its emphasis as HRM managers were expected to be more strategically focused, proactive, and process based in order to contribute to organisational success. Changing demographics and increasing shortages of workers with the needed capabilities also began growing in importance.

Contemporary Challenges for Organisations and HRM

In recent years increasing attention has been paid to the importance of HRM in determining an organisation's competitive advantage. As noted earlier, some have even declared that human resources represent the only enduring source of competitive advantage available to many of today's organisations. A number of factors have contributed to the increased attention on the value of HRM. For example, there are a number of changes in organisations themselves and broader trends causing these changes to occur. Perhaps most importantly, organisations today are under intense pressure to be better, faster, and more competitive. There are more and more efforts to squeeze productivity out of some organisations, while others are merging and downsizing. Why is this the case? Technological changes, deregulation, and globalization are three trends accounting for these competitive pressures. Other trends include diversity, workforce changes, and achieving societal goals through organisations. Let's take a brief look at each of these.

Technological Changes

Technology has been forcing and enabling organisations to become more competitive. Technology is also changing the nature of work. For example, telecommunications already makes it relatively easy for many to work at home, and the use of computer-aided design/computer-aided manufacturing (CAD/CAM) systems plus robotics is booming. Manufacturing advances like these will continue to eliminate blue-collar jobs, replacing them with jobs requiring greater skill, and these new workers will require a degree of training and commitment that their parents never dreamed of. As a result, to remain competitive, jobs and organisational charts will have to continue to be redesigned, new incentive and compensation plans instituted, new job descriptions written, and new employee selection, evaluation, and training programs instituted—all with the help of HRM.

Deregulation

Being better, faster, and more competitive is also more important because for many industries the comfortable protection provided by government regulations continues to be stripped away. In the United States (and in many other industrialized countries such as England, France, and Japan), industries from airlines to banks to utilities must now compete nationally and internationally without the protection of government-regulated prices and entry tariffs.

One big consequence of deregulation has been the sudden and dramatic opening of various markets. One need only look at the long-distance phone companies' efforts to enter the previously protected monopoly of companies like AT&T or the efforts of startups in the airline industry to compete head-to-head with industry giants like Delta and American Airlines. Just as significant has been the impact that deregulation—and the resulting new competition—has had on prices, requiring these organisations to get and stay "lean and mean." Prices for hundreds of services like long-distance calls have dropped in some instances, which means organisations must get their costs down, too.

Globalization

One of the most dramatic challenges facing U.S. organisations as we begin the twenty-first century is how to compete against foreign firms, both domestically and abroad. Many U.S. companies are already being compelled to think globally, something that does not come easily to organisations long accustomed to doing business in a large and expanding domestic market with minimal foreign competition. The Internet is fuelling globalization, and most large organisations are actively involved in manufacturing overseas, international joint ventures, or collaboration with foreign organisations on specific projects.

Such globalization has vastly increased global competition. The implications of a global economy on HRM are many. From boosting the productivity of a global labor force to formulating compensation policies for expatriate employees, managing globalization and its effects on competitiveness will thus present major HRM challenges in the years to come.

Workforce Diversity

Managers across the United States are confronted almost daily with the increasing diversity of the workforce. The workforce is continuing to become more diverse as women, minority group members, and older workers flood the workforce. More specifically, one need only note that in many large urban centres, such as Miami, New York, and Los Angeles, the workforce is already at least half composed of minorities. Women with children under age six have also been one of the fastest-growing segments of the workforce. The HRM function will increasingly be called upon to help organisations accommodate these employees with new HRM programs and with basic skills training where such training is required.

As the workforce gets older employees will also likely remain in the workforce well past the age at which their parents retired, due to Social Security and Medicare funding shortfalls and the termination of traditional benefit plans by many employers. Increased diversity presents both a significant change and a real opportunity for managers. Organisations that formulate and implement HRM strategies that capitalize on employee diversity are more likely to survive and prosper.

Changes in the Nature of Work

Technological changes, deregulation, and globalization are also changing the nature of jobs and work. For one thing, there has been a pronounced shift from manufacturing jobs to service jobs in North America and Western Europe. As the number of manufacturing jobs have decreased over the past two decades, the number of part-time and service industry jobs in fast foods, retailing, legal work, teaching, and consulting has increased. These service jobs will in turn require what have recently been referred to as "knowledge" workers and new HRM methods to manage them as well as a new focus on human capital.

The knowledge, education, training, skills, and expertise of an organisation's workers make up an organisation's human capital. And human capital is more important than it has ever been before. Service jobs put a bigger premium on worker education and knowledge than do traditional manufacturing jobs. Even entry-level factory jobs have become more demanding. For example, factory jobs in the textile, auto, rubber, and steel industries are being replaced by knowledge-intensive high-tech manufacturing in such industries as telecommunications, pharmaceuticals, medical instruments, aerospace, computers, and home electronics, and even heavy manufacturing jobs are becoming more high tech. Human capital is quickly replacing machines as the basis for most organisations' success.

An important realisation for managers and organisations is that these new "knowledge" workers can't be just ordered around and closely monitored like their parents. New HRM systems and skills will be required to select, train, and motivate such employees and to win their commitment.

Changing Employee Expectations

As levels of education have increased within the population, values and expectations among employees have shifted. There has been a steady increase in the number of employees with a college degree. The result has been an emphasis on the increased participation by employees at all levels. Previous notions about managerial authority continue to give way to employee influence and involvement, along with mechanisms for upward communication and due process.

Another expectation of employees is that the electronics and telecommunications revolutions will improve the quality of work life. Innovations in communications and computer technology will accelerate the pace of change and as a result lead to many innovations in HRM.

Finally, organisations are taking steps to support employees' family responsibilities as the proportion of dual-career families continues to increase along with the number of women in the workforce. More and more companies are introducing family-friendly programs that give them a competitive advantage in the labor market. Once able to assume that the demands of male employees' home lives were taken care of by their wives, organisations are now being pushed to pay attention to family issues such as day care, sick children, elder care, schooling, and so forth. One result is that more employees are enjoying the opportunity to work at home. These pro grams reflect HRM tactics that organisations use to hire and retain the best-qualified employees, male or female, and they often pay off.

Managerial Changes

Management approaches are adding to the challenges facing HRM. Empowerment of employees and self-managed teams are two specific management approaches that are having a significant impact on today's HRM function. Empowerment is a form of decentralization that involves giving employees substantial authority to make decisions. Under empowerment, managers express confidence in the ability of employees to perform their work. Employees are also encouraged to accept responsibility for their work. In many organisations now using self-managed teams, groups of employees do not report to a single manager; rather, groups of peers are responsible for a particular area or task. The breadth of changes in areas like managerial challenges continues to have a powerful impact on today's HRM personnel.

Chapter 8

A Study of Accessing Data Warehouses Using the Internet and Intranets

Today, accessing data warehouses has been made easier for a company's decision makers because of the pervasive influence of the Internet and the World Wide Web. The Internet has opened the door to a global business community where being able to make informed decisions and move quickly is important. Following on the heels of the Internet are intranets which are private networks that leverage the Internet's infrastructure to extend corporate client/server networks to users wherever they may be. Currently, companies are leveraging the Internet's infrastructure to extend their corporate networks. The reasons are clear: intranets are easy to use, rapidly implemented, cost-effective, and efficient for making information and knowledge available to the people who need it. The fact is that any authorized user can get to information on an intranet with nothing more than a standard web browser.

Essentially, the web offers a real information and knowledge highway to companies who want to extend data warehouse access to others in their supply chain (i.e., suppliers, distributors, and even customers). Web browsers are cheaper than most client tools. And maintenance releases are much simplier to administer since updating the client application on the web simply requires updating the server. The bottom line is that the web offers a compelling new way to deliver data warehousing applications that can broaden the reach and value of data warehousing investments.

Until recently, intranets were basically stores of large volumes of information, much of which was static corporate communications materials, such as telephone listings, internal memos, corporate communiques, and similar documents. Today, because intranets have become more pervasive, the kinds of information and knowledge on them has changed and expanded. Companies use their intranets to give users dynamic access to the information and knowledge in their databases and data warehouse that can help them improve the speed and quality of their decisions.

An important goal of a data warehousing strategy that includes the Internet and a company's intranets is to ensure that its information and the knowledge derived from it has value and that a company and its clients and suppliers can use it when and where it is needed. For the typical data warehousing manager, web-based distribution reduces the complexity and delay associated with supporting remote users. New web development tools automatically link web pages to databases or data warehouses. Among the key tools are those based on Sun Microsystems' JAVA language, which is embedded in the latest version of Netscape; Oracle's web-enabled tools; and web-enabled ROLAP (relational on-line analytical processing) tools from Dimensional Insight, Information Advantage, and MicroStrategy. Because both webs and data warehouses have a common goal of data access, newer releases of query tools will be web-enabled so that web-browser software is able to access data warehouses. In fact, innovative user companies like MasterCard are already building such interfaces between their applications. MasterCard now offers banks access to its customers' buying habits, for instance, based on their credit card transactions.

Problems with Data Warehouses

Just as with data marts, there can be problems with data warehouses. For one, there needs to be support from top management that centres on an enterprise approach to data warehousing. If there is no all encompassing solution, but rather one that addresses only one part of the business, the data warehouse will never be fully embraced or utilised by the entire organisation. From this view, the company has no clear vision of the end-state data warehouse. For real success, the goals of the data warehouse program must be aligned with the company's strategic goals. Although this may seem obvious to the reader, this may not be obvious in everyday corporate life.

Another typical problem that a company faces is a technology-based approach to data warehousing. There is a need to take a close look at users' fundamental needs to be sure they are seeking highly innovative and strategic capabilities. Such goals demand a basic change in design and development that, if ignored, is likely to result in less than optimum results. Other problems lie not in the technology but in the decision making that determines the uses, expectations, and means of data warehousing applications.

More specifically, they include the lack of leadership to get the data warehouse up and running, poorly understood relationships between technical capabilities and business benefits, absence of executive input regarding data warehousing applications, poor strategic use of data and information to arrive at important knowledge about the company, and lack of up-front overall involvement by computer personnel in the computer department.

Still other problems with data warehousing centre on data quality and performance. Many data warehouses had to be substantially redesigned due to data quality and performance issues. Generally, there is a need to assign several analysts and a manager who will maintain the warehouse's data integrity and performance over time. As data warehouses grow in number as well as in size, it is necessary to create and maintain a data repository road map for users. A data repository road map enables users who look at the data warehouse to track where the data and information originated. Additionally, data warehouse developers should provide the capability to maintain software and hardware independence. That is, there should be the ability to select from a variety of hardware and software programs to ensure that a company has the flexibility to switch to best-of-breed tools for data access and/or data mining.

Typical Data Warehousing Software

Before focusing on representative data warehousing products per se, it is wise to state that most data warehousing software products currently focus on one of these three areas:

(1) acquisition,

(2) storage, and

(3) access.

Most software vendors that provide these products have a proven track record for performing one of these functions well. Generally, most have simply retrofitted existing products to meet warehousing requirements. In the first category, there are a number of vendors who have products designed to manage and automate the acquisition process. Some of the data acquisition product vendors have begun to integrate several of their products into complete warehouse offerings.

The largest vendors in this category include Prism Solutions Inc., Carleton Corporation, and Platinum Technology.

In the second category, there are a number of storage software products. Currently, firms like Oracle, Sybase, IBM, and Informix control the bulk of the market. When a typical Unix server or mainframe computer seems incapable of handling the envisioned workload, MIS executives look to SMP (symmetric multiprocessing) or MPP (massive parallel processing). For the truly large jobs, SMP may not be enough. In those cases, some companies are turning to MPP machines. Although SMP machines can handle up to 16 additional processors at a time, MPP systems incorporate the use of dozens or even hundreds of processors. The power of these machines cannot be approached by any other hardware on the market today.

In the third category, access software products include the following: First, unlike the overly simplistic report writers of the past, query facilities tools turn a large, complex data-warehouse environment into a friendly, well-managed workstation. Second, access software tools centre on statistical analysis as found in products like SAS and SPSS. Third, a number of data discovery products are getting a lot of attention. These include decision support tools as well as artificial intelligence and expert systems. Using neural networks, fuzzy logic, decision trees, and other advanced mathematical and statistical tools, these products allow users to sift through massive amounts of raw data to discover new, insightful and, in many cases, useful things about the company, its operations, and its markets. These data discovery products are widely used in data mining as covered later in the chapter.

In addition to the three areas above, representative data warehouse software typically includes data visualization and on-line analytical processing. Data visualization represents a graphical rendering of information from data warehouses. As such, data visualization software products bring graphical representation to new heights. A popular visualization tool that falls under this group is geographical systems. Such systems turn data about stores, individuals, or anything else into easy-to-understand, dynamic maps.

On-line analytical processing (OLAP) centres on multidimensional analytical tools for accessing, storing, and manipulating decision support and EIS (executive information system) style information. These tools represent a whole new generation of high-powered, user-friendly data investigation systems that allow users to look at information from numerous different perspectives. OLAP software products provide the capability to slice and dice reports dynamically, and to look at the same kinds of information from different perspectives. For example, at the press of a button, a product manager is able to view sales figures for a given product at the national level, or view them broken down by division, or drill down to view figures by territories within a division, or check sales numbers for each store in a territory and then compare them against sales of stores from a different territory. This last category of data access products can enrich a decision maker's capability to ask and answer important questions facing him or her on a daily basis.

Utilisation of Data Replication in Data Warehousing

In the past, multiple corporate databases were consistently synchronized and were, in essence, clones of one another. This task, frequently called "nightly refresh," has been practiced for years in the world of computer mainframes. When a few PCs were added, the job progressed into downloading and uploading data between the PCs and the mainframe. The situation was quite controllable. However, when there is a network, servers, users spread across multiple time zones, groupware applications, and reliance by users on real-time information, this chore grows into a network manager's nightmare, better known as replication.

Essentially, replication copies information from one database to another so that the contents of both databases are consistent. That can mean transferring data from a central mainframe to branch servers and then down to local workstations, from news feeds to reporting and analysis applications, between networked servers, or within almost any architecture. Data movement can be one-way or bidirectional. It can be event driven (triggered by data value changes) or time dependent (performed at regular intervals or nightly).

Typically, there are many benefits to replication. It is often implemented to offload processing onto a single server. Copies of corporate data are down-loaded to branch offices where departmental users can access the local data more efficiently. Replication can also be an important part of a data warehouse strategy—that is, consolidating data from multiple operational databases to a single data store for analysis. Having multiple copies of data also sets the stage for quick disaster recovery and cost-efficient load balancing on busy networks.

Replicated DBMSs are recommended for applications such as backup as well as knowledge management, OLAP, and DSS systems that do not require up-to-the-minute information and knowledge. Most managers who use these systems do not require up-to-the-minute facts.

Most companies doing backup do not require the up-to-date capabilities of two-phase commit. That is, many users would probably prefer working with a backup system that was slightly out of sync to waiting for the main database management system to be restored. Replication also allows companies to divide up a database and ship information and knowledge closer to those users who work with it the most. Response time improves because the information and knowledge is stored locally rather than at a central site. Wide area network costs fall because users no longer need to access the network to work with what they need.

Virtual Data Warehousing

From a different perspective, companies may find that building a data warehouse may be more than they are able to undertake at the present time. A vendor such as INTERSOLV allows companies to migrate to storage-focused and subject-area databases as needed. The end product is just-in-time data warehousing that is a rapid approach to providing infrastructure transparency for companies while enabling immediate end user access to enterprise information and knowledge. Overall, a virtual data warehouse allows a company to operate a data storage facility that can be implemented at a fraction of the time and expense required for a data warehouse.

INTERSOLV's Virtual Data Warehouse looks like its Q+E end user, data access tool, somewhat enhanced and renamed Explorer. There is no distinct physical data warehouse since the data resides in the production systems. The virtual warehouse organises and smoothes out access to the data for users. This approach is similar to those taken by end user access tools such as Business Objects' Business Objects and Dimensional Insight CrossTarget. Basically, INTERSOLV is providing an open semantic mapping layer. It is this mapping layer, which INTERSOLV calls SmartData, that disguises the complexity of the multiple back-end data sources and allows users to access, for example, "Sales" without knowing from where the sales data is actually coming. Other products also provide a semantic mapping layer, but they are proprietary. INTERSOLV, a developer of open database connectivity (ODBC) drivers, employs industry standard ODBC as the access method.

INTERSOLV's Virtual Data Warehouse establishes access to the data first, without requiring heavy analysis and infrastructure. Decision makers can access information and knowledge from any data source in everyday business terms, working with the presentation tools of their choice. Typically, virtual data warehousing is most useful in small-scale situations, since it allows users to get up and run quickly with data access as an interim step to building a real data warehouse. Most of the experts concede that the virtual data warehouse will not replace the real data warehouse.

Data Mining or Knowledge Discovery

Data mining, also known as knowledge discovery, explores the knowledge held within a company's database by revealing patterns and trends that can suggest improved performance in terms of greater customer satisfaction, higher quality products, savings, and profits. In effect, data mining uncovers hidden patterns and provides predictive trends which can be easily applied to benefit the business. Since the raw materials for data mining are abundant, data contains records that could potentially reveal hidden patterns and predictive trends that could further a company's mission, its objectives, and its measurable goals. Sales records, for example, could reveal highly profitable retail sales patterns. In turn, financial analysts could find future trends that are highly profitable. As another example, an engineering firm could determine the combination of data conditions (e.g. manufacturing time, lot size, assembly parameters, operator number, reject rate, etc.) that determine the quality of the products being produced. In the final analysis, the key to successful data mining is to employ those tools that result in real knowledge discovery.

To discover meaningful patterns in a data set, consider, for example, gross margins that are stored in a retail sales database. Gross margins fluctuate over the course of a year with a 10 percent increase between summer and autumn. A company executive might be tempted to conclude that sales margins generally increase from summer to autumn and that the increase in gross margin depends upon the season. In reality, there are many other potential variables housed in the database that could influence gross margin. These might include quantity sold, discount rate, commission paid, customer location, other purchases made, and length of time as a customer. If the discount rate is greater in the summer than in the autumn, the increase in gross margin could simply be a result of a lower discount rate and have nothing to do with a change in season.

By removing the effect of discount rates on gross margins, gross margins might be found to be higher in the summer. In the final analysis, data mining examines all potential explanatory factors and associated data elements to ensure that the best pattern is retrieved from the data and that no potentially misleading effects are introduced into the chosen patterns. Going one step further, in place of showing the effect of only one condition, such as season, on gross margin, data mining can show the combined effect of a pattern, such as a particular time, location, and discount rate, that produces the maximum gross margin. By replicating that pattern, a longer-term strategy that will systematically increase gross margin and associated profitability can be established. That optimal pattern becomes a basis for predicting future trends.

Tie-In of Data Mining with Data Warehousing and OLAP Systems

Today, data mining is considered to be a direct outgrowth of the data warehouse concept. It allows data access and analysis that gives users unparalleled capabilities to extract hard-to-get-at data, spot trends, and recognise patterns in corporate databases. While many companies have built data warehouses to consolidate data located in disparate databases, they are implementing data mining to learn more about the data. Although data warehouses are a part of the technology mainstream, available data mining tools now allow companies to improve their knowledge of their customers and markets. For example, retailers want to use data mining to target customers for sales promotions better or to manage inventory across various geographic locations. Similarly, telecommunications companies want to use it to forecast demand patterns, profile and segment customer groups, customize billing, and analyze profitability. In addition, financial services companies are employing data mining to consolidate information from multiple sources, analyze customers' business patterns, and sell them more services.

Today, data warehousing and data mining have the following requirements in common: They need very large databases and highly scrubbed and integrated data sources, along with facts that provide insight and new directions for further analysis. Even though these items are complementary, data warehousing is a prerequisite for data mining.

Typically, the business objective is to extract patterns, trends, and rules that are implicit within data warehouses to evaluate proposed business strategies. In turn, a company can improve its competitiveness, return on investment, and business processes. Data mining utilises a number of traditional technologies, such as RDBMS, massively parallel processing (MPP), symmetric multiprocessing (SMP), and statistics as well as newer technologies, to provide pattern recognition and analysis, including neural networks, inductive reasoning/decision trees, and advanced data visualization. Essentially, data mining is built on the foundation laid by a successful data warehouse. Most decision support solutions require an effective data warehouse in combination with multiple data mining and OLAP tools, in addition to high-end systems integration assistance to pull it all together for decision makers.

As just noted, most data analysis is done with data mining and OLAP tools. Data mining tools help users find answers to questions that they never thought to ask. They do this by characterizing the patterns inherent in the data, developing a hypothesis from these patterns, and using the hypothesis to predict future behaviour. In contrast, OLAP allows users to explore corporate data from a number of different perspectives. Using drill-down analysis, users can discover the facts and implications of a high-level trend, or discover individual implications. OLAP servers and clients provide an easy-to-use multidimensional interface for drilling down from high-level to low-level summary data and for slicing and dicing data. They can also provide support for complex derived functions that are common in financial reporting and consolidation. Some of these tools are based on a multidimensional database (MDD) while others impose a multidimensional model directly on relational data. OLAP tools have powerful analytic capabilities and are well suited for fast response time, complex calculations, and budgeting/forecasting "what-if" analysis.

Inasmuch as on-line analytical processing provides top-down, query driven data analysis, data mining provides bottom-up, discovery driven analysis. Data mining identifies facts or conclusions based on patterns discovered. OLAP, and in particular data mining, are two emerging data analysis tools that impact the typical organisation's use of corporate knowledge assets. In essence, these tools are reinventing the way companies think about business capabilities, customer satisfaction, and strategic goal-setting for the future.

Types of Data Mining Information

Because data mining reaches much deeper into databases, its tools provide users with the capability to find patterns in data and infer rules from them. Those patterns and rules can be used to guide decision making and forecast the effect of those decisions. In addition, data mining can speed analysis by focusing attention on the most important variables. Although users might find these patterns with a series of queries against the data, data mining lets the users explore a much wider range of possibilities than even the most sophisticated set of queries.

Typical types of information that can be obtained by data mining include forecasting, associations, sequences, classifications, and clusters. Essentially, all of these may involve predictions. However, the first type (forecasting) is a different form of prediction in that it estimates the future value of continuous variables—like sales figures—

based on patterns within the data. The second type of information centres on associations that happen when occurrences are linked in a single event. For example, a study of supermarket baskets might reveal that when potato chips are purchased, 60 percent of the time some type of cola is also purchased—unless there is a promotion, in which case cola is purchased 80 percent of the time. Knowing this fact, managers can evaluate the profitability of a promotion.

In the third type of information or sequences, events are linked over time. For example, if a house is bought, then 60 percent of the time a new oven will be bought within one month and 65 percent of the time a new refrigerator will be bought within three weeks. For the fourth type, or classifications, patterns are recognised that describe the group to which an item belongs. It does this by examining existing items that already have been classified and inferring a set of rules. Because a common problem for many companies is the loss of steady customers, classification can help discover the characteristics of customers who are likely to leave and provide a model that can be used to predict who they are. It can also help a company determine what kinds of promotions have been effective in keeping which types of customers. The fifth and last type of information is clustering. It is related to classification, but differs in that no groups have yet been defined. Using clustering, the data mining tool discovers different groupings within the data. This can be applied to problems as diverse as detecting defects in manufacturing or finding affinity groups for bank cards.

Past Approaches to Multidimensional Knowledge Discovery

Several approaches over the years have been utilised in multidimensional knowledge discovery. Essentially, all try to use the information contained in many fields in the data store to explain variations in the outcome field (i.e., the field in question, such as profit, defects, interest rate, and so forth). One of the earliest methods used is multiple regression, which is a standard statistical technique that uncovers the pattern of dependencies between multiple predictor fields and the outcome. Another method is decision tree analysis, which was developed to compensate for some problems that had arisen from using multiple regression as a data mining tool. Decision trees show the combined dependencies between multiple predictors and the outcome as a number of decision branches.

The user is able to see how the outcome changes with different values in the predictors. More recently, neural networks have been used. They resemble multiple regressions in many ways, except that rather than using statistical theory as the basis of the technique, they imitate the information processing methods of the human brain.

In reference to decision trees, J. N. Morgan and J. A. Sonquist were interested in developing a statistical analysis technique that would more accurately depict "real world" social and economic events than was possible using standard statistical regression techniques. They proposed an "automatic interaction detector," which is an idea based on earlier notions of designing a data discovery tool that was modelled on the techniques used by humans to solve problems.

They suggested a mechanical or automatic system that would mimic the steps taken by an experienced data analyst to determine any strong data interaction effects. An interaction is the kind of effect that was observed above among discount rate, season, and gross margin. The automatic interaction detector technique involves an exhaustive examination of all possible relationships between predictor fields and the outcome field to determine the strongest, or best, prediction. When found, the data is divided into two groups that are determined by the predictor field, and the process is repeated for the descendent groups that are formed by the selection of the strongest predictor. The result is a series of splits, or branches, in the data. Each split produces a new data set that is in turn split. The final result is a series of branches that resemble a decision tree.

When the Morgan and Sonquist procedure was implemented as a company program called AID (Automatic Interaction Detector) at the University of Michigan in the early 1970s, it was enthusiastically received by many researchers, statisticians, and data analysts. The beauty of decision trees and of the original AID program is that they address the problem of hidden dependencies and spurious relationships. Decision trees provide models that are easier to interpret than those produced by the mathematically intense regression and neural network techniques. Furthermore, given the way a decision tree is built, the analyst can more easily control the model's construction and can assemble a more valid and reliable final result.

In reality, observers found some problems with AID. For one, AID was overly aggressive in identifying relationships and dependencies in the data. Furthermore it could not discriminate meaningful relationships from meaningless relationships. Often, therefore, AID would select relationships that were the result of random fluctuations in the data set. As a result, these decision trees frequently did not reflect the actual data patterns that affected the question under examination.

In the mid-1970s, G. V. Kass published a method of addressing the shortcomings of AID. His approach applied the lessons of statistical hypothesis testing to the decision trees produced by AID. Kass reasoned that the branches identified by AID could be tested by using a standard statistical test to determine whether they were the result of a chance effect of fluctuations in the data. He found a way to apply modern statistics to determine the statistical strength of each of the decision tree branches. When the branch was too weak statistically, it would not be displayed. In the early 1980s, a validation approach emerged to further remedy the problems with the AID technique. This approach complements Kass's statistical approach because although statistical approaches are based on statistical theory, validation approaches are based on an examination of the actual properties of the data. Validation looks closely at the data used to build the decision tree. If too many fluctuations are observed, then as in the statistical approach employed by Kass, the branch is rated poorly and is not presented in the final decision tree display. This validation approach is also used in a modern decision tree product called KnowledgeSEEKER, marketed by Angoss Software International.

Current Data Mining Methods

A starting point for data mining methods and software available from vendors is a brief introduction to how data mining works. Typically, a decision maker feeds a business goal and corporate data into data mining tools, which then use a variety of methodologies to model the data. The outcome is a set of factors in the data that are related to, or predictive of, the business goal in the form of data visualization. Using data visualization software that presents a picture for users to see, an enormous amount of information is presented in a concise format. A very wide range of information presented to the user, including knowledge, allows peaks or valleys to stand out.

Many methods currently used in data mining are natural extensions and generalizations of analytical methods that have been used for decades. Neural networks, a special case of projection pursuit regression, were developed in the 1940s. CART (classification and regression trees) methods were used by social scientists in the 1960s. K-nearest neighbour, a form of density estimation, has been used for a half-century. Other methods include rule induction, discriminant analysis, and logistic regression. Basically, these methods, just like regression techniques, model relationships between a set of profile variables and an outcome. However, what is new is that these methods are now being applied to more general business problems, caused by the increased availability of data, inexpensive processing power, and user-friendly software for implementing these methods. The recent interest in data mining is in part due to the improved user interfaces. Varieties of regression techniques, discriminant analysis, and even simple graphs can help reveal hidden patterns. Because no single method solves all or even a majority of problems, successful data mining requires a portfolio of tools, both old and new.

Networking & Telecommunication

This new program, the Technology Opportunities Program, or TOP, was initially known as the Telecommunications and Information Infrastructure Assistance Program or TIIAP. It was designed to provide matching grants to community-based organisations including community development groups; state, local and tribal governments; health care providers; school districts; libraries; universities; public safety services; and other non-profit organisations to help them access new opportunities through telecommunications. TOP projects form a sub-set of what are increasingly being called community technology or community informatics projects.

TOP is housed in the National Telecommunications and Information Administration (NTIA), an agency of the U.S. Department of Commerce. NTIA is the Executive Branch's principal voice on domestic and international telecommunications and information technology issues. NTIA works to spur innovation, encourage competition, help create jobs and provide consumers with more choices and better quality telecommunications products and services at lower prices.

This chapter provides an overview of TOP's mission and strategy. It discusses key characteristics, model projects, and lessons learned in three categories related to community development: rural resource development, urban asset mapping, and community economic development. It includes a summary of the results of program-wide evaluations conducted periodically since the program started. The chapter concludes with a section that discusses elements that successful projects have in common, for the use of community development practitioners who are considering using ICTs to amplify, support, or augment their efforts.

Top Mission and Strategy

TOP's mission is to promote the widespread availability and use of ICTs in the public and non-profit sector. In terms of strategy, TOP is a national program that emphasizes innovation, learning, diffusion of new ideas, and practical knowledge. TOP projects yield new insights into how best to use ICTs, provide concrete examples of best practices, and stimulate innovation in communities across the country. What works well in Montana or St. Paul might also work well in Alabama or Spokane.

TOP requires grantees to focus on measurable outcomes throughout the life span of each project. Once a project is funded, the grantee works with TOP staff to develop outcomes and milestones against which they are evaluated on a quarterly basis. TOP requires all grant recipients to conduct independent evaluations of their projects. TOP expects that ongoing evaluation activities will lead to continual improvement, assessment of short-term impacts, and development of indicators of long-range impact. Taken collectively, these evaluations provide practitioners with guidance in the development of new projects. Individual project evaluations provide TOP with information for program planning and design. TOP actively shares selected evaluations and what grantees are learning in general through outreach activities and through the TOP Website.

TOP's goal is to demonstrate how non-profit and public sector organisations can use ICTs to increase productivity, create new jobs, help educate children, and provide better medical care to all Americans, in addition to bringing the benefits of new ICTs to traditionally unserved and under-served groups. Aside from TOP's own evaluation studies, discussed in more detail below, these outcomes have been well documented in the literature since the mid 1990s, as summarised in the next section of this chapter. From 1994-2004, TOP made 610 grants in all 50 states, Washington, D.C., the U.S. Virgin Islands, and Puerto Rico, granting approximately $233.5 million in federal funds, matched by local communities that provide $313.7 million. TOP projects cut across many different sectors including community and economic development; lifelong learning and the arts; health; public safety; and social, governmental, and other public services.

Although all of TOP's projects use technology as a tool, the program's focus is on how the technology can be used to make a positive difference in communities. The criteria for applying for these grants include a project's purpose, innovation, community involvement, evaluation, dissemination, feasibility, and budget. The proposed project's future sustainability—whether the project will continue after the grant period ends and once community acceptance has been realised—is a consideration as well. Most important, the reviewers look for the new knowledge that can be adapted in other communities around the country and developed through implementation of the project.

The dynamics and potential for positive results through collaborative community development efforts have been discussed widely in the literature. TOP grantees are required to work in partnership with others in their community to carryout their projects. These collaborations must be documented in the application and throughout implementation of the project. Many TOP grantees have reported that significant and sustainable benefits have occurred in their communities while putting together the partnerships even during the application process.

It is a bit difficult to review the literature of community technology projects, as they are dual in nature. They focus on technology on one hand, but take place in a number of different community sectors, i.e., community development, arts and heritage, education, health, public safety, government, and so forth. As a result, the field has a vast bibliography. Some research as taken place on projects that use technology as a tool to make positive change in communities, yet each field in which these projects take place has its own literature. Reviewing these individual bodies of knowledge is far beyond the scope of this chapter. Instead, it will summarise a narrower subset of resources that attempt to chronicle the use of technology as a tool for community development efforts in general. Please note that the citations in this section are not meant to be exhaustive, but they illustrate some of the work that has taken place in this field.

In the widest sense, this field has its theoretical basis in social network and complexity theory. Many students and researchers of community technology projects point to a resonance between community technology projects and the work of complexity theorists such as Waldrop, Kelly, Kauffman, and Holland; and writers on social and technological networks including Granovetter, Castells, Putnam, Barabasi, and Lawrence Lessig. Valdiz Krebs is doing interesting work on mapping social networks, and has published white papers on his web site. Most community networking practitioners see their efforts within the framework of asset-based development theory.

Several Internet researchers have described the history of community technology projects. As early as 1993, Steve Cisler, a librarian with Apple Computer, wrote a history of projects that he had interacted with since the early 1970s, entitled "Community Computer Networks: Building Electronic Greenbelts." The following year he convened the first national conference to focus on these projects and expanded on this history in an address entitled "Community Networks: Past and Present Thoughts." These treatments are interesting in their discussion of community technology projects just before the explosion of graphical access to the World Wide Web.

That year also saw the publication of Howard Rheingold's The Virtual Community: Homesteading on the Virtual Frontier, which he revised and re-issued in 2000. This is a vivid description of virtual communities including one of the first national electronic forums, The Well. In 1995, Anne Beamish wrote her thesis at MIT entitled "Communities On line: A Study of Community Based Computer Networks." Gary Chapman, currently Director of the 21st Century Project at the LBJ School of Public Affairs at the University of Texas has written often and well about community technology projects from their very early days. His publications are listed in his online biography.

In 1994, the Benton Foundation began to focus their efforts around digital divide issues. They hosted a number of important meetings and published widely on the topic. Their Web site remains an important resource for the field. Most recently, they published A Broadband World: The Promise of Advanced Services and E-government for All: Results of National Survey.

Most recently, he completed Community Technology Centres as Catalysts for Community Change for the Ford Foundation. In 2003, Gurstein and others held a colloquium resulting in the formation of the Community Informatics Research Network, and a statement for the World Summit on the Information Society.

In 2003, Using Community Informatics to Transform Regions edited by Stewart Marshall and Wal Taylor was published, which looks at community informatics as social capital in various contexts around the world. Howard Rheingold's Smart Mobs: The Next Social Revolution discussed how wireless mobile devices such as PDAs and cellular telephones, or "cell phones," are shaping our patterns of interaction. In addition, in 2004, Anthony Wilhelm, director of TOP, published Digital Nation: Toward an Inclusive Information Society. The Benton Foundation and the Association for Public Technology are currently updating their 2003 report on the promise of broadband A Broadband World: The Promise of Advanced Services. The Community Technology Review and The Journal of Community Informatics are the primary periodicals in the field.

Virtually all of TOP's projects are involved in community development in one way or another, because of TOP's focus on problem, solution, and outcome. Each grantee must articulate a compelling problem in their community, devise a creative and innovative solution using ICTs, and discuss how they plan to measure the impact of their efforts in their community. In many cases, community development organisations have been grantees or partners in these projects. A search of TOP's database of funded projects using the keywords "community development" brings up almost every project funded since 1994.

The following section summarises characteristics, provides a case study, and discusses lessons learned in three community technology project categories: rural resource management, urban asset mapping, and community economic development. Additional areas in which TOP has invested can be explored via TOP's database of grants on the TOP Web site. These areas include arts and culture, education, lifelong learning, workforce development, telemedicine and telehealth, social services, human services, and public safety.

These particular cases were selected following a review of all projects funded by TOP during the last ten years in these three areas, on the occasion of a request to present lessons learned at the 2003 annual meeting of CTCnet. Program Officers scanned final reports in order to select projects that were good examples of type, and to yield compelling lessons learned. The selected projects are illustrative of three very different kinds of TOP projects, whereas the lessons learned and elements for success described in Sections V and VI are derived from the full spectrum of TOP projects.

Rural Resource Management

Almost half of the projects TOP has supported since 1994 have taken place in—or have partners in—rural parts of the country. Technology projects in rural communities face a host of unique issues. First, there are unique economic issues. In rural areas, the exploitation of resources and absentee ownership produce profits that do not remain in the community, which causes widespread poverty. There is often a parallel problem with out-migration, especially of young people. On the other hand, rural areas are increasingly viewed as desirable places to live, and new technologies have made it easier for people who can telecommute to move to rural areas. Unfortunately, over the past few years, investment in the kind of information infrastructure that makes this possible has slowed in the rural United States.

Second, rural regions have issues of access. Access is made difficult simply by the physical distances involved. In some places, there is still no phone service available or, if there is, it is very poor. Or communities might have a number of very small telephone companies with confusing service areas and/or duplication of administrative infrastructure. Finally, some rural populations, including tribes, exhibit a cultural resistance to the adoption of technology and a tendency toward conservatism.

The rural projects that TOP has funded through the years mirror the diversity of rural communities themselves. For example, TOP funded projects to provide Internet service where there was none in rural communities above the Arctic Circle; rural health initiatives in

Nebraska and South Dakota; and projects designed to revitalize faltering rural economies in the Appalachian coalfields. TOP funded projects focusing on rural transportation in eastern Maine and rural housing in Kentucky. The Oklahoma First Project created online weather prediction for rural Oklahoma. TOP supported projects to increase law enforcement capacity in rural Colorado and emergency response time in rural Oregon.

Project Snapshot: Sevier River Water Users Association

The rural western United States, in particular, is characterized by a high degree of concern about water management. Even in the not-so-distant past, people have been killed in water disputes. In recent years, drought has exacerbated the struggle for power—who gets water when and who manages the water system. In the past, building dams and irrigation canals was the solution to these problems. Now, environmental concerns make new construction unlikely; therefore, better management of the existing resource is essential. The Sevier River Water Users' Association in Utah developed an interactive water management network to improve water management and water conservation.

The project employs a solar, spread-spectrum radio communications system, environmental sensors, and computers to create an integrated water management system using remote wireless/ telephone transmission of data from monitoring stations. Users can see graphically, as well as visually, through streaming output from solar powered cameras exactly what is happening in the watershed. The system enables water orders to be placed over the Internet and provides real-time data to canal companies, irrigators, county residents, and farmers. The project resulted in improved water management and conservation, and better crop yields throughout the Sevier River Basin. The TOP project allowed this region to avoid the costly construction of a new dam on the river.

Lessons Learned

Several aspects of the Sevier River project illustrate how ICTs can be especially significant in rural areas. In areas where there are few resources, it is important that everyone involved comes to the table. In this case, the project was an opportunity for various parties, which were once at conflict about use of a scarce resource, to create an environment of complete transparency, in which all information is available to everyone, all the time. Collaboration was essential to planning and carrying out the project, and equal access to data led to an environment more conducive to the development of new collaborative projects.

The project has transformed the way participants think about working together. Whereas in the past, community members thought in terms of product, now they think in terms of process. The Project Director noted, "With 'everybody-get-involved' style development, the product evolves over time in concert with technological change and maturing water user needs. As prototypes (both hardware and software) are rushed to the field, feedback is critical. It becomes necessary for everybody involved in the project to interact, something the Internet facilitates". The Sevier Water Project is an excellent example of a project that overcame rural distances using ICTs. Whereas before a ditch rider had to ride or walk the whole length of the canal system and report back, often with already out-of-date information, this system affords instant and timely data to all users, which also made decision making faster and more effective.

This project struggled with technological obstacles. They found that the value of the data on the Web sites was enhanced by improving Internet access. Water managers made much more use of the system when they had high speed, full-time, broadband connections. The water managers, who were provided with low-cost, license-free spread spectrum radio links, accessed the Web resource more frequently than others did, sometimes twenty times per day. Now project staff is thinking about enhancing the system with the addition of telephone touch-tone data retrieval for cell phone and wireless personal digital assistant access.

The Sevier River Project demonstrates that technology is insensitive to those who can use it. Project-generated data ended up being just as useful to tourists who boat on the river as it was to the farmers who were the designated beneficiaries. This particular project was so successful that it provoked interest in adaptation as far away as China, where the project team consulted with engineers on the Yellow River Dam project.

Urban Asset Mapping

Deploying technology projects in urban areas presents another set of challenges. Inner cities, especially, share difficult economic crises, like capital flight, stagnant growth, and out-migration with rural areas, but these problems take place in a much more dense and complex field. One of the most persistent urban challenges is affordable housing; and the larger problem of managing neighbourhood and community assets is another. The very complexity of the urban environment makes sharing information through ICTs a powerful tool for capacity building and regeneration.

A landmark TOP-supported project that is now being deployed statewide and nationwide started out in Los Angeles. Neighbourhood Knowledge Los Angeles (NKLA) is a prime example of how information technology can be deployed to address a complex set of significant social, economic, and cultural issues in a tangled urban environment. NKLA developed from a planning grant TOP awarded in 1996 to the Resident Controlled Housing Association. The Association developed a plan to increase access to information as a way of combating urban decline. They started from a very simple but powerful assumption: the process of urban decay often begins with small, little-noticed changes.

In 1998, TOP awarded a second grant to the Advanced Policy Institute at the University of California at Los Angeles to implement the plan. Researchers collected information on tax delinquencies, building code violations, unpaid utility bills, and other variables from disparate government databases, amassing them into a single, Internet-accessible database that serves as an early warning system for neighbourhood activists and government officials combating urban problems.

From this simple premise—that community residents, armed with tools that enhance their ability to understand the complex forces affecting their communities, can transform information into a tool to promote and realise community objectives—NKLA expanded to include community asset mapping, programs for young people, projects to empower disabled Angelinos, work with the city's Housing Crisis Task Force, and development of a countywide information system. The project is currently expanding further to become the statewide Neighbourhood Knowledge California project, and it is working with the Fannie Mae Foundation on national deployment.

Through its creative use of state-of-the-art information strategies and its focus on providing users with sophisticated but easy-to-use online tools, NKLA became a national model, indeed an international model, of how community leaders and residents of low-income neighbourhoods can use information tools to strengthen their communities. In addition, tools that were once considered cutting-edge are now considered, through the example of NKLA, as basic strategies—e.g., online mapping of community assets, constantly updated statistics, and community online forums.

Several key lessons were learned in implementing the original citywide project. First, the Web can be used to provide access—and give new meaning—to previously difficult-to-obtain data. Second, interactive mapping on the Web allows for data integration and layering; and third, the Web provides an excellent method for decentralizing data collection and building the technical capacity of community residents. Using GIS technologies, NKLA has been able to integrate data resources in a way that would have been virtually impossible using other means.

Like the rural example, these project participants also note that projects like NKLA can open up space for new forms of participation and decision-making. The transparency of information leads to new opportunities to collaborate and make decisions together. Building the technical capacity of community residents not only results in more effective community action, but also transfers to other aspects of residents' work, including research and public presentations.

Participants learned that content is essential. Part of the success of the project was its rigorous emphasis on high quality and relevant data presented in a highly usable fashion. In addition, participants found that social networks and relationships are critical for the success of community technology projects. Participants discovered that regular face-to-face interaction was critical in the development and the operation of NKLA.

Furthermore, they found that a Web site alone can only do so much to improve neighbourhoods. A Web site, no matter how sophisticated, robust, and compelling, is not THE answer to community problems, but it can be a vital and even indispensable tool. Community groups can use this tool to more effectively monitor and analyze neighbourhood problems and design more creative planning solutions armed with the kind of information provided by NKLA.

The other important lesson learned through projects as NKLA is the necessity for flexibility. The project team's capacity to adapt to new conditions and the mechanisms built into NKLA for determining user satisfaction with the system, have made it possible for the project director to improve the project continually. In addition, this flexibility has enabled him to demonstrate the potential of NKLA for other communities in California, as well as to such diverse locales as Brazil and Sweden. The "one-stop" data access concept that is now considered conventional wisdom in community networking projects was, in large part, developed and demonstrated by NKLA.

NKLA continues to define the state-of-the-art in integrating public policy and online technologies, and it has become one among a growing number of TOP-supported initiatives deploying sophisticated information tools to address social policy issues, such as urban decline, the scarcity of affordable housing, and disparities in homeownership. The Cleveland Housing Network, Right Moves Net in Chicago, Neighbourhood Information Exchange in New York, and a homeownership project designed by the National Council of La Raza also address these issues.

Community Economic Development

Advanced telecommunications can be both a catalyst for the development of many kinds of businesses and a market niche for the development of tech-specific businesses. In addition, telecommunications becomes increasingly important as economic development increasingly takes place in a regional context. Since its inception, TOP has supported at least a hundred projects that focus on economic development. Some projects, including a grant provided to the State of West Virginia, focus on using ICTs to market communities to prospective industries. Another group of projects focuses on welfare to work, including a project in Boston designed to improve business efficiencies for inner city daycare providers. The Northern Indiana Workforce Investment Board is using ICTs to improve workforce skills, and projects in East Baton Rouge and Columbus, Ohio, provide online links to employment opportunities. Others, such as the Minnesota Rural Partners' Biz Pathways, is designed to connect small business service providers with each other and new clients

TOP funded a number of projects designed to provide a broad spectrum of supportive services for new business development. The Philadelphia Enterprise Centre is one such project. TOP supported projects that explored technology as a market niche for new business development. For example, ACEnet's project in Appalachian Ohio trained high school students to start and run their own technology consulting businesses to meet the needs of adult entrepreneurs for computerized bookkeeping, marketing materials, and so forth.

Project Snapshot: The Accelerator Online

The Accelerator Online Project focused on the twin problems of high unemployment and low-income levels in largely rural areas of central California. At the time of the proposal, California was experiencing an economic boom; however, low-income rural job seekers—many of whom spent long hours in seasonal agricultural work, with poor public transportation options—were unable to take advantage of training and business programs offered through California State University (CSU). In response, the University Business Centre of CSU-Fresno made courses more accessible through Web-based, interactive distance learning. In addition, project partners used distance-learning technology to teach entrepreneurship to 225 high school seniors at Central Valley High School.

An important component of the project was the development of an online incubator for start-up companies. The companies received computers, domain names, and access to e-commerce servers. In addition, they received valuable consulting services, as well as access to an MBA internship program; an outsourcing databank; and online and classroom training on how to raise money, determine product feasibility, develop a marketing plan, and boost their accounting skills. The Fresno Hispanic Chamber of Commerce translated the course materials into Spanish, while the Hmong community chose to take the courses in English in order to improve their English language skills.

The Accelerator Online Project had significant impact on the community, fostering the creation of a number of unusual small businesses. One participant in the project created a business that designs greeting cards for families of people who are incarcerated, employing incarcerated artists to create the unique greeting card illustrations. Using Accelerator Online and a computer at one of the project computer labs, this entrepreneur was able to craft a business plan and present it to potential investors.

One of the most significant lessons learned in the Accelerator Online project was that partnerships were at times challenging because of partner-staff turnover and differences in commitment level to the project. However, eventually the Accelerator Online project increased the number of participating partners by facilitating provision of equipment, staff, and learning resources in exchange for community access, space, local contacts, and promotional and marketing activities. They learned to screen partners more carefully for both records of accomplishment and commitment before initiating partnerships.

Another lesson offered was the need for projects that support entrepreneurship to be especially flexible in eras of economic and market volatility. Certainly, the dot.com boom and bust has demonstrated the importance of flexibility and good business planning. In an environment characterized by constant change, they found that a focus on participant feedback and continuous improvement was vital to meeting the needs of their clients.

Project participants also note the challenge of trying to use technology to jumpstart entrepreneurship when the technology is changing so quickly. They advise using existing technology—i.e., tested and stable technology—rather than trying to invent new technology. The innovation should rest in the training that is delivered, not the delivery mechanism. They met with technological challenges in terms of lack of connectivity and lack of computer skills. These basic problems need to be addressed from the outset, before programmatic material is delivered.

These grantees learned some valuable lessons about course design. Their most successful classes were highly interactive, and they featured blended learning. Courses promised and delivered tangible outcomes. In addition, the classes themselves were flexible enough to change to accommodate the needs and wishes of the students.

Most economic development projects have to take a systemic approach to putting people back to work. These projects have to include ways to overcome the many barriers to employment or business ownership, including the cost of child care, transportation, and acquiring workplace skills. For example, another successful TOP grantee offers micro-enterprise and workplace skills training as well as computer literacy classes for students starting their own businesses and learning how to be excellent employees. They have a fleet of vans that pick these students up at their homes for classes, and they provide free childcare while students are learning.

Evaluation

In February 1999, NTIA released its first evaluation report of TOP grantees for the years 1994-1995. This study examined short term project impacts and examined the potential for long term impact. The study methodology included a comprehensive review of applications and progress reports, a survey of 206 projects, and site visits to 24 projects. The evaluators concluded, "The 1994 and 1995 TIIAP projects helped change the way in which millions of end users and other beneficiaries access information and services".

The 1999 study discussed ways to improve community technology projects. The evaluators recommended that all projects include a comprehensive community needs assessment. Less successful projects experienced problems when they failed to develop a long-term vision of how ICTs would eventually benefit the community. Second, the evaluators suggested that recipients be required to collect outcome data. Both of these suggestions were incorporated into later grant rounds and TOP policy for grantees. The TOP Web site provided assistance in offering examples of exemplary projects that developed long-term vision and suggestions on how to collect outcome data.

In 1999, NTIA released thirty-six case studies of individual projects. Each case study described how each project was implemented, what was accomplished, the impact on the community, lessons learned, and how project staff envisioned future plans. The case studies report addressed trends uncovered across the rural and urban sites, and steps taken by projects to sustain the project after the grant-funding period.

The collected case study report identified three trends that were occurring across rural and urban projects. First, in most sites, partners were active in project implementation. In fact, the boundary between partner and end-user often became blurred in these projects. Second, the projects used a variety of techniques to involve and engage stakeholders, including establishing advisory boards, conducting regular project meetings, and conducting needs assessment activities. Third, training models across the projects fell into three categories: stand-alone sessions, ongoing series, and train-the-trainer. Most sites employed more than one mode. The training was most beneficial when it was tailored to the needs, learning style, and availability of its target audience.

The report addressed the problems grantees face most frequently which are the time, cost, and effort required to develop technology infrastructure; underutilisation because of lack of community involvement; stakeholder and end-user buy-in; staff turnover; problems with partner relations; and underestimation of time.

In terms of sustainability, the evaluators found that TOP grantees from those years used three strategies to ensure their projects would continue after the federal investment was finished: private or public grants/donations, fee structures, and the absorption of project-related costs. Most projects employed at least two of these. For example, one grantee encouraged their partners to build costs of technology into the budget for all aspects of their programs, thereby spreading the costs over a number of budget categories, rather than leaving them a line item by themselves. In the words of the respondent, "We are not asking foundations to fund computers for a literacy class. We are asking them to fund a literacy program that includes access to a computer".

The case study report concluded with five best practices that can increase the likelihood that projects will be sustained. These are suggestions for community technology practitioners to conduct a needs assessment before even starting to assure that the technology will be used by the intended beneficiaries; to carry out a feasibility study in advance to assure that the technology can be used as planned; to involve stakeholders and partners in planning and operations; to integrate technology in ongoing activities, so that it is seen as a means to an end instead of an end unto itself; and to collect data that can be used to demonstrate and publicize the projects benefits.

A second study conducted in 2000 surveyed 49 of the projects funded by TOP in 1996. The purpose of this study was to assess the effects of the projects at the local and national levels. These evaluators found that a wide variety of organisations served as grant recipients.

Recipients worked with multiple partners, and the primary contribution of the partners involved human resources. In terms of project implementation, this study showed that the vast majority of the projects were designed to address multiple barriers to using telecommunications technologies, and they used a wide variety of strategies to increase this access. Most of projects were successful, and they were more rigorous in their data collection than the earlier grantees. Insufficient planning continued to pose the greatest challenge to successful implementation.

As with the 1994 and 1995 projects, nearly all of the 1996 grantees were still in operation at the time of the survey. They found that increased user base, financial contributions from partners, and partner buy-in were factors in continuation and growth. Over half of the projects had expanded to serve additional users. Johnson & Johnson Associates conducted the most recent evaluation study cleared for public consumption in 2001.

The report summarised findings related to forty-two TOP projects funded during 1996 and 1997 that are not covered in the 2000 report. These evaluators found that TOP projects continued to improve services, provide learning and training opportunities, and remove barriers to the use of new ICTs. TOP grantees implemented a wide variety of activities, and most generated spin-off activities not initially envisioned in the grant application. The most common obstacle grantees experienced had to do with underestimating time for planning, inadequate or under-qualified staffing, or lack of commitment and follow-through. A large majority of the projects reported that TOP funding was imperative to the projects' success, and they were still in operation after the funding period ended.

Trends in Community ICt Projects

Over the past decade, the problems that grantees have chosen to tackle tended to become increasingly focused. Early in the program, for example, TOP funded a number of multi-purpose community networks—projects that simply provided Internet access and a Web presence where there was none in places like rural western North Carolina and remote rural Alaska. TOP funded generic early-distance education projects in western states and developed some early telemedicine applications. At that time when there were still many communities with no access to the Internet, and when distance education and telemedicine were in their infancy, it made sense for TOP to support these efforts.

By contrast, present-day grantees tend to use a much wider spectrum of ICTs to address very specific problems. For example, in the last cohort of grantees, the Gulf of Maine Aquarium is creating a collaborative model for environmental monitoring using wireless handheld technology to collect and share aquatic information in Maine. The University of Pittsburgh is creating portalware that makes Web sites more accessible for people with blindness and low vision. The Lane County, Oregon, Council of Governments is creating a computerized simulation that will make it possible for them to map inefficiencies in a complex criminal justice system in order to increase effectiveness in the system. The Missoula Demonstration Project has created an online bank for advanced directives and healthcare powers of attorney so that these living wills are more easily available whenever and wherever they are needed.

In the last few years, TOP applications have increasingly focused on how wireless broadband might be deployed to provide last mile connectivity to rural and remote places that still do not have access. In addition, several recent grantees are experimenting with interactive transactional technologies, like those employed by commercial Web sites such as Amazon or Web-based applications like Turbo tax.

For example, both the Benefits Check Up project and Medicare Rights project developed personalized interactive online toolkits to help people access customized information on their Medicare benefits. They have consulted together on their projects. Another grantee, Seedco in New York City, is helping the newly employed apply online for the benefits for which they still qualify. Action for Boston Community Development has created an online interface to streamline billing between childcare providers and the State of Massachusetts.

Other technology trends include the availability of streaming audio and video, making it possible for dancers to collaborate in real time from studios in Ohio and Minnesota; and for young Vermont artists to perform for their distant e-mentors. For the last several years, the TOP grant round has attracted increasingly sophisticated use of Geopositioning and Geographical Information Systems technologies, particularly in the area of public safety. For example, the Right Route project in Oregon provides hand-held computers to first responders, giving them accurate directions and interactive maps to any address in the region. The last TOP grant round included successful applications experimenting with new technologies including Wi-Max and Broadband Over Power lines.

Finally, powerful demographic trends are reflected in recent TOP applications by the increased need for translation into languages other than English for information provided over the Internet. In Lincoln, Nebraska, immigrants and refugees can log into the Lincoln Action Program's Website in eight languages.

Common Elements for Success

Successful TOP projects shared two key characteristics. First, successful TOP projects become as proficient at building strong social networks as they are at building effective digital networks. Second, successful TOP grantees were realistic about the amount of time it takes to put together, implement, and measure the impact of these complex projects. The lesson most frequently articulated by TOP grantees is this: "The people part takes a lot longer than the technology part." Technology projects are sufficiently complex that collaboration is key to success. Thus, people skills are crucial to successful implementation. Partnerships are required to cover the costs of technology projects and to manage the project. The most successful TOP projects involved all the stakeholders in the initial design of the project. Everyone worked harder when they were involved in deciding what happens, and this inclusiveness held true for partners, advisory board members, staff, volunteers, and end-users

It is especially crucial to identify and enlist the enthusiastic participation of local leaders. This participation can be complicated by distance in rural areas and by the complexity of leadership networks in urban areas. Once local leaders embrace the project and begin to participate, it is crucial to continue to involve them throughout the life of the project. Community involvement is one factor that may have nothing to do with the technology but everything to do with the project's eventual success. In almost every case that a TOP project has not succeeded, it was because of lack of community buy-in and ongoing participation.

Collaborative projects by definition are political. TOP grantees frequently cite political challenges, including staff turnover at partner agencies and organisations and the impact of elections as key problems or factors causing delay in implementation. In the last few years especially, projects that had state governments as partners had to be especially creative in securing their match, since state budgets were reduced. In addition, government entities often have different attitudes about timelines and deliverables than non-profit organisations. Partnerships between organisations with different cultures, in particular, needed to be very carefully defined and fully committed, and then carefully nurtured. Many successful grantees recommend formalizing these relationships in at least memoranda of understanding, if not contractual agreements.

Another challenge that TOP grantees encountered over and over again was the challenge of working with technology vendors. Often, project staff has little experience in the process of producing vendor specifications, detailing criteria against which bids will be evaluated and so forth. Grantees encountered lengthy project delays because of insufficient care in vendor selection and vendor turnover.

An associated need is for adequate documentation. The problems caused by turnover can be ameliorated by meticulous documentation of every feature of system policy, procedure, and operations. Successful grantees required vendors to document how the software and hardware works, and how it is maintained in language that can be understood by non-technical personnel on staff. Finally, in working with vendors, it was essential to define intellectual property rights carefully. All kinds of headaches—not to speak of possible lawsuits—can be avoided through careful definition of ownership from the outset.

TOP grantees repeatedly cited the importance of recruiting the appropriate people and maintaining their level of enthusiasm. Staff must be maintained just as rigorously as equipment and partnerships. Opportunities for training and professional development are significant. Mechanisms for motivating volunteers are especially important since they receive no monetary incentive to continue to participate.

The second lesson most grantees will say they learned is that you cannot plan enough, test enough, and build in too much time for planning and testing. Grantees frequently underestimate the time and effort it takes to carry out a project, the time it takes vendors to accomplish their work, the need for technical support, and the time involved in developing and delivering training. They tend to overestimate the reliability of the technology. And they find it is more difficult than expected to secure local acceptance, and to work with local telcos and ISPs.

In addition, many grantees in general have experienced problems with scope. When planning a technology project it is important to know what is achievable, to keep expectations under control and avoid allowing them to blur reality, and to understand when objectives need further refinement or limitations. Successful grantees developed a detailed project plan with activities and milestones that guaranteed that the project was reflected and continually improved upon a sit moved along. Practitioners need to build ongoing technical support into their sustainability plans. In every community technology project, the technology often seemed to change faster than it can be installed, and once installed, it required maintenance and care throughout the life of the project. In conclusion, the three primary qualities of successful TOP grantees are commitment, patience, and persistence, both in terms of working in collaboration with others, and having the foresight to plan and the flexibility to change that plan as necessary.

Internet and Electronic Commerce

Before we explore the legal and commercial realities of e-commerce it is first necessary to consider the nature of the Internet. We must do this in order to be able to apply the law to it. Common misconceptions can arise merely from the description of the Internet. Use of 'the' implies that it is some constant 'thing' which has definable edges and a shape and even a legal presence. Actually the Internet is a communications infrastructure, nothing more. It is made up of countless thousands of computers that are connected together by means of telecommunications systems. Not all these computers will be connected at any one time and hence it is impossible to define its size and presence. When one user's computer is in communication with another user's computer, the information that is sent and received by each will travel in different and unpredictable routes and will pass through many computers on its journey. All the computers and communications services providers that are not either the original sender or the final recipient of the information are intermediaries or 'conduits'.

Internet connectivity is commonly established by using an Internet Service Provider (ISP). ISP's are organisations that have a permanent presence on the Internet and provide a fixed line or dial-up service. Some ISP's make a charge for this service, but many do not. However, most businesses will require a website, and hence web space on the ISP's server, and this is normally a chargeable service. Businesses wishing to establish a web presence for the first time are advised to consult the trade press (business Internet magazines such as The Industry Standard) for information on ISP's.

The Nature of E-Commerce

Before looking at the legal detail and common pitfalls of e-commerce, we will initially consider the nature of e-commerce — what exactly is it? We will also look at some terminology that is frequently used by e-commerce lawyers and business people, always bearing in mind that what we are dealing with here is not the technology, but rather the way the law applies to the transactions that are undertaken using the technology. E-commerce (or electronic commerce) is the term used to denote a commercial (usually contractual) transaction that takes place between two or more people using the communications infrastructure known as the Internet. The first thing to note is that there is a range of e-commerce transactions, the following of which are examples:

- the purchase by an individual of a book from a commercial website such as that run by Amazon.com
- the ordering by Company A of office stationery from Company B's website
- the exchange of e-mails between two persons whereby it is agreed that a certain service will be carried out by one party in exchange for a fee from the other.

We do not usually think of the latter type of transaction as falling within the category of e-commerce, but it is just as much an electronically formed contract as the other two examples: an exchange of e-mails can result in a legally binding contract. For example, if I send you an e-mail offering to sell you one dozen doughnuts, and you respond accepting my offer, this is just as much an e-commerce transaction as one taking place using the colourful pages and flash technology of the World Wide Web.

Initially it will be important for an e-commerce business to identify whether its e-commerce transactions will be classified as B2B (business-to-business) or B2C (business-to-consumer). This is because the law applies in a different way to each. Of course, it may be that the e-commerce business anticipates engaging in both types of transaction. The differences between these two types of transaction will be referred to throughout this Report. Other terms commonly referred to throughout this Report are as follows:

- World Wide Web — that part of the Internet which uses the Hyper Text Transfer Protocol (HTTP) to display so-called web pages and to allow links between such pages anywhere on the Internet.
- ISP (Internet Service Provider) — a provider of Internet connectivity.
- ASP (Applications Service Provider) — an online provider of computer applications, such as software.

The subject of e-mails and the potential liabilities that can arise for employers from their inappropriate use is a subject in its own right and will not be considered in detail in this Report.

The Law of Commerce

The law does not currently recognise an e-commerce transaction as being inherently different from a non e-commerce transaction. However, due to the online nature of e-commerce, particular aspects of the law apply differently where a transaction takes place online. This section will briefly consider those aspects of the law that apply to both online and offline transactions. This will be followed by an introduction to those areas of the law that have particular application in the case of e-commerce.

None of the topics listed below — aspects of law that apply to commercial transactions generally — will be looked at in any detail in this Report, but it is important to put the e-commerce transaction in the context of its commercial and legal setting.

Trading Vehicle

In the case of an existing business wishing to move online, the legal infrastructure will already be in place. For start-ups it is important to consider the way that trading is to be carried out. It may be that the business will be a partnership. Alternatively it might be a registered limited company. E-commerce start-ups should usually be advised to use a company for trading purposes. The advantages are many, the two main ones being that it is easier to attract investment capital, and that the owners of the company (its shareholders) will benefit from limited liability.

Shares and Responsibilities

This is an issue for start-ups. If a company is chosen as the appropriate trading vehicle then there will be a number of issues that must be agreed upon by the participants: What are the appropriate proportions for share ownership? Who are the directors going to be? Who will be Company Secretary? Will a shareholder agreement be needed? Appropriate legal advice should be sought on these issues.

Taxation

Once again the normal considerations will apply to both offline and online transactions. The position is slightly complicated in an e-commerce transaction due to the greater likelihood that such a transaction will involve parties who are not located in the same country.

Intellectual Property Rights

The subject of intellectual property is relevant to all businesses but may be particularly important to e-commerce as the business model may involve a greater use of intellectual property.

Terms and Conditions of Trade

Again, this is relevant to all businesses. The terms on which a business obtains its supplies, and upon which it trades with its customers, are of fundamental importance. Incorporation of a business's terms into its contracts is not usually a problem with offline transactions but significant issues commonly arise with an e-commerce business. Choice of law and jurisdiction will be an issue in some contractual arrangements, both on and offline.

The Law of E-Commerce

Of greater significance, and the subject of this Report, are those areas of the law that apply specifically to e-commerce. This may be due to the fact that 'old' law applies in an unusual way to e-commerce due to the unpredictable (at the time that law was made) nature of the transaction. Or it may be due to the presence of 'new' law, which, though it may not have been formulated to deal with e-commerce exclusively, clearly has it in contemplation as the main target of its intended regulation. The list that appears below contains those areas of law that are of particular (and in some cases exclusive) interest to e-commerce businesses:

Formation of Contract

How does the law of offer and acceptance apply to an online transaction? In other words, how will a contract be formed in the virtual world? Will a web page that displays a product for sale be an offer or an invitation to treat? This issue has yet to be decided by the courts.

Much of the value in an idea for an e-commerce website can be tied up in the Intellectual Property Rights (IPR's).It is essential, particularly when trying to attract seed money from venture capitalists, that these IPR's are delineated and transferred to the trading vehicle. They may be owned by a number of different people and so IPR assignments will need to be executed at an early stage. In any event the value of IPR's is notoriously underestimated by businessmen. Appropriate steps must be taken to protect this value and to take steps to safeguard any brand name and associated goodwill.

Copyright

Copyright is of course comprised in IPR's, but there are some specific considerations in relation to copyright that need to be looked at. We know that the law of copyright protects literary and artistic works and that these are the sorts of works that comprise a web page. So far so good. But what about links from one web page to another? Could they constitute breach of copyright? Of course in most instances, there would be no complaint because a link from one website to another would generate increased traffic and this surely is the objective of websites. But in one case a newspaper's website contained a link to the news section of another newspaper's website. It was held in an interim application to the court that this could constitute copyright infringement. The reason for the case was that the link bypassed the front (or home) page of the other website and so the users did not get to see the advertising messages on that home page — this technique is known as deep-linking. Sites that undertake deep-linking should be aware of the potential legal challenges that could arise.

Domain Names and Cybersquatting

The fact that every website must have a unique Internet address means that many commercial enterprises will be disappointed in their desire for a specific domain name. This is the problem that is inherent in a first-come-first-served system of domain name registrations. Difficulties can arise when one business feels it should be entitled to use a domain name that has already been registered by someone else.

Data Protection

This area of the law governs what may and what may not be done with individuals' personal information. In many cases much of the value in an e-commerce business is in its customer database, which may consist of a variety of information including name, address, e-mail address, date of birth, shopping habits, annual household income, etc. The law establishes a code of conduct (known as the Eight Data Protection Principles) for the processing of such data as well as a right for every individual to see a copy of such data if they request it.

Distance Selling Regulations

Towards the end of 2000, and in response to European Union legislation, the UK passed the Consumer Protection (Distance Selling) Regulations. These regulations require all UK businesses that enter contracts with consumers 'at a distance' (clearly this includes, but is not exclusive to, Internet transactions) to do two main things that they were not legally obliged to do before. The first is to provide certain specific information to the consumer and the second is to allow consumers a 'cooling-off' period of seven working days from receipt of goods to return the goods for a full refund.

IT Requirements

There will be a number of legal considerations which arise from the infrastructure and services which are required for an e-commerce business. Hence there may be purchase or rental of hardware and software, rental of server space if required and a website hosting and development agreement where the hosting is to be outsourced. These topics are considered where relevant throughout this Report.

Law and Jurisdiction

One of the difficulties with an Internet transaction is that the buyer and seller may be in different parts of the world. The question then arises as to which legal system will govern the contract in the absence of any binding express provision. A related question but one which is more complex and somewhat political is which court system will have jurisdiction to hear any relevant litigation?

Information Systems for Business Operations

For all sized organisations, change will be a constant in the next millennium. Technological advances, including non-technical ones, are coming more quickly than ever before, thereby affecting all aspects of a typical company. As network computing technology progresses, there are increasing pressures to create a new type of workplace. The current workplace will experience a transition that will fragment it into myriad on-site and off-site offices. Not only will the work environment change dramatically in the 21st century, but there will also be a need for business systems to support the new work environments.

More specifically, there will be a need for new types of systems that focus on discovering knowledge that responds to the changing environment. Although past and current information systems have served the needs of management and operating personnel, newer types of business systems, such as those stressed in this text (i.e., knowledge management systems [KMSs]), will be at the forefront of these newer types of systems found in a typical company. An important goal of knowledge management systems is to provide competitive advantage by giving decision makers (from the highest level to the lowest level) the necessary insight into patterns and trends that affect their domain. In effect, a broad-based KMS environment challenges decision makers to evaluate changing times more thoroughly. Such an environment allows decision makers to tailor their information and related knowledge requirements by discriminating according to user-defined criteria.

Essentially, a knowledge management system is capable of making comparisons, analyzing trends, and presenting historical and current knowledge. But more importantly, such a system enables decision makers to analyze and understand the patterns quickly and identify the most significant trends. As such, it is an accurate predictive method for decision makers.

In addition, a knowledge management system can track and evaluate key critical success factors for decision makers, which is valuable in assessing whether or not the organisation is meeting its corporate objectives and goals. Overall, a knowledge management system can assist decision makers in making better informed decisions that affect all aspects of a company's operations.

As will be seen in this chapter and future ones, past management information systems basically used the computer as a means of providing information to solve recurring operational problems. A better approach today, however, is the utilisation of knowledge management systems to position decision makers at the centre of the decision-making process. By increasing the capabilities of decision makers, a KMS environment improves the chances that an organisation will achieve its goals of increased sales, higher profits, and so forth by placing knowledge and related information in the hands of decision makers at the proper time and place and by providing flexibility in their choice and sequence of analysis and in the final presentation of results. From this enlarged perspective, knowledge management systems provide essential knowledge and related information to decision makers so that they can better cope with changing times. Additionally, there is a tie-in between knowledge management systems and past and current management information systems, with emphasis on OLAP (on-line analytical processing) systems, expert systems, neural networks, and virtual reality systems.

Bibliography

Aarseth, E. J.: *Cybertext: Perspectives on Ergodic Literature*, Baltimore: Johns Hopkins University Press, 1997.

Allen, T., and M.S. Morton: *Information Technology and the Corporation of the 1990s*, Oxford University Press, New York, 1994.

Beatty, K.: *Teaching and Researching Computer-Assisted Language Learning Longman*, London: Ungar Press, 2003.

Bromley, H.: *Education/Technology/Power: Educational Computing as a Social Practice*, Albany: Suny Press, 1998.

Christopher, K.: *Introduction to Educational Technology*, New York: Columbia University Press, 2001,

Clark, Robert G.: *Comparative Programming Languages*, Addison-Wesley, England, 1988.

Cryer, J.D., & Cobb, G.W.: *An Electronic Companion to Business Statistics*, NY: Cognito Learning Media, Inc, 1997.

Disessa, A. A.: *Changing Minds: Computers, Learning, and Literacy*, Boston: MIT Press, 2000.

Dytham, C.: *Choosing and Using Statistics: A Biologist's Guide*, Boston: MA: Blackwell Publishing, 2003.

Ernest, B.: *National Science Education Standards: Observe, Interact, Change, learn*, London: Edward University Press, 1998.

Fill, C., :. *Marketing Communications. Contexts, Contents and Strategies.* London, UK: Prentice Hall Europe, 1995.

Flynn, Roger R.: *Information Science,* New York: Marcel Dekker, Inc., 1987.

Gabriel, Richard P.: *Patterns of Software: Tales from the Software Community*, New York: Oxford University Press, 1996.

Gupta, B.M.: *Libraries, Archives and Information Technology: An Annotated Bibliography, 1970–1990,*Aditya Prakashan, New Delhi, 1991.

Hammer, R. M. Hocks, U. Kulisch, D. Ratz: *C++ Toolbox for Verified Computing I. Basic Numerical Problems,* Berlin Heidelberg, Springer, 1995.

Hancock, J.: *Teaching Literacy Using Information Technology,* Newark: International Reading Association, 1999.

Iversen, G.R.: *Bayesian Statistical Inference (Quantitative Applications in the Social Sciences),* Newbury Park, CA: Sage Publications, 1984.

James, E.: *Meaningful Learning using Technology: What Educators Need to Know,* New York: Oxford University Press.1997.

John, P.: *Information Technology and Authentic Learning: Realising,* London: George Allen & Unwin Press, 1998.

Kelly, D.: *Technology-Enhanced Learning: Principles and Products,* Madison: University Wisconsin Press, 1996.

Lawrence L.: *Integrating Information Technology into the Teacher Education,* Buckingham: Open University Press, 1995.

Marylou Hale: *New Automation Technology for Acquisitions and Collection Development,* New York: Haworth Press, 1995.

Mattfeldt, H.: *Enhancing the Learning Experience in Economics and Institute for Learning and Research Technology,* New York: Wiley Press, 2001.

Michael, M.: *Technology Education for Teachers,* Lexington, KY: University Press of Kentucky, 1996.

Neumaier, A.: *Interval Methods for Systems of Equations,* Cambridge University Press, Cambridge, 1990.

Philip, G.: *Learning with Technology: A Constructivist Perspective,* UK : Addison-Wesley Press, 1995.

Robert,D.: *21st Century Skills: Learning for Life in Our Times,* Baltimore, MA: John Hopkins University Press, 1997.

Rowley, J.E.: *Organising Knowledge: an Introduction to Information Retrieval,* Gower, Aldershot, 1987.

Rudy, B.: *Technology, Literacy and Learning: A Multimodal Approach,* London: Palgrave Macmillan, 1978.

Salaberry, M.R.: *The use of technology for second language learning and teaching: a retrospective,* The Modern Language Journal, 2001.

Schmidt, S. J. : *Werbung, Medien und Kultur,* Westdeutscher Verlag, 1995.

Seitel, Fraser P.: *The Practice of Public Relations,* Macmillan, Delhi, 1994.

Vogt, W.P.: *Dictionary of Statistics and Methodology: A Nontechnical Guide for the Social Sciences,* CA: Sage Publications, 2005.

Webster, Frank, and Robins, Kevin: *Information Technology—A Luddite Analysis,* Ablex, Norwood, NJ, 1986.

Williamson, Judith : *Decoding Advertisements,* Marion Boyars Publishers Ltd., 1994.

Index

❑❑❑